Privacy Vulnerabilities and Data Security Challenges in the IoT

Internet of Everything (IoE): Security and Privacy Paradigm

Series Editors:
Vijender Kumar Solanki, Raghvendra Kumar, and Le Hoang Son

IOT
Security and Privacy Paradigm
Edited by Souvik Pal, Vicente Garcia Diaz and Dac-Nhuong Le

Smart Innovation of Web of Things
Edited by Vijender Kumar Solanki, Raghvendra Kumar and Le Hoang Son

Big Data, IoT, and Machine Learning
Tools and Applications
Rashmi Agrawal, Marcin Paprzycki, and Neha Gupta

Internet of Everything and Big Data
Major Challenges in Smart Cities
Edited by Salah-ddine Krit, Mohamed Elhoseny, Valentina Emilia Balas, Rachid Benlamri, and Marius M. Balas

Bitcoin and Blockchain
History and Current Applications
Edited by Sandeep Kumar Panda, Ahmed A. Elngar, Valentina Emilia Balas, and Mohammed Kayed

Privacy Vulnerabilities and Data Security Challenges in the IoT
Edited by Shivani Agarwal, Sandhya Makkar, and Tran Duc Tan

Handbook of IoT and Blockchain
Methods, Solutions, and Recent Advancements
Edited by Brojo Kishore Mishra, Sanjay Kumar Kuanar, Sheng-Lung Peng, and Daniel D. Dasig, Jr.

Blockchain Technology
Fundamentals, Applications, and Case Studies
Edited by E Golden Julie, J. Jesu Vedha Nayahi, and Noor Zaman Jhanjhi

Data Security in Internet of Things Based RFID and WSN Systems Applications
Edited by Rohit Sharma, Rajendra Prasad Mahapatra, and Korhan Cengiz

Securing IoT and Big Data
Next Generation Intelligence
Edited by Vijayalakshmi Saravanan, Anpalagan Alagan, T. Poongodi, and Firoz Khan

Distributed Artificial Intelligence
A Modern Approach
Edited by Satya Prakash Yadav, Dharmendra Prasad Mahato, and Nguyen Thi Dieu Linh

Security and Trust Issues in Internet of Things
Blockchain to the Rescue
Edited by Sudhir Kumar Sharma, Bharat Bhushan, and Bhuvan Unhelkar

Internet of Medical Things
Paradigm of Wearable Devices
Edited by Manuel N. Cardona, Vijender Kumar Solanki, and Cecilia García Cena

For more information about this series, please visit: https://www.crcpress.com/Internet-of-Everything-IoE-Security-and-Privacy-Paradigm/book-series/CRCIOESPP

Privacy Vulnerabilities and Data Security Challenges in the IoT

Edited by
Shivani Agarwal, Sandhya Makkar,
and Duc-Tan Tran

CRC Press is an imprint of the
Taylor & Francis Group, an **informa** business

First edition published 2021
by CRC Press
6000 Broken Sound Parkway NW, Suite 300, Boca Raton, FL 33487-2742

and by CRC Press
2 Park Square, Milton Park, Abingdon, Oxon, OX14 4RN

© 2021 Taylor & Francis Group, LLC

CRC Press is an imprint of Taylor & Francis Group, LLC

Reasonable efforts have been made to publish reliable data and information, but the author and publisher cannot assume responsibility for the validity of all materials or the consequences of their use. The authors and publishers have attempted to trace the copyright holders of all material reproduced in this publication and apologize to copyright holders if permission to publish in this form has not been obtained. If any copyright material has not been acknowledged please write and let us know so we may rectify in any future reprint.

Except as permitted under U.S. Copyright Law, no part of this book may be reprinted, reproduced, transmitted, or utilized in any form by any electronic, mechanical, or other means, now known or hereafter invented, including photocopying, microfilming, and recording, or in any information storage or retrieval system, without written permission from the publishers.

For permission to photocopy or use material electronically from this work, access www.copyright.com or contact the Copyright Clearance Center, Inc. (CCC), 222 Rosewood Drive, Danvers, MA 01923, 978-750-8400. For works that are not available on CCC please contact mpkbookspermissions@tandf.co.uk

Trademark notice: Product or corporate names may be trademarks or registered trademarks, and are used only for identification and explanation without intent to infringe.

Library of Congress Cataloging-in-Publication Data

Names: Agarwal, Shivani, editor. | Makkar, Sandhya, editor. | Tran, Duc-Tan, editor.
Title: Privacy vulnerabilities and data security challenges in the IoT / edited by Shivani Agarwal, Sandhya Makkar, and Duc-Tan Tran.
Description: First edition. | Boca Raton, FL : CRC Press, [2021] | Series: Internet of everything (IoE). Security and privacy paradigm | Includes bibliographical references and index.
Identifiers: LCCN 2020020017 (print) | LCCN 2020020018 (ebook) | ISBN 9780367334390 (hardback) | ISBN 9780429322969 (ebook)
Subjects: LCSH: Internet of things--Security measures.
Classification: LCC TK5105.8857 .P754 2021 (print) | LCC TK5105.8857 (ebook) | DDC 005.8--dc23
LC record available at https://lccn.loc.gov/2020020017
LC ebook record available at https://lccn.loc.gov/2020020018

ISBN: 978-0-367-33439-0 (hbk)
ISBN: 978-0-429-32296-9 (ebk)

Typeset in Times
by Deanta Global Publishing Services, Chennai, India

This book, Privacy Vulnerabilities and Data Challenges in the IoT, *is dedicated to the memory of my beloved father, Sh. Shiv Avtar Jindal. He is my light, my rock, my friend, my confidant, my protector, my provider, my guide, my father, my mother, and my hero—he is my everything.*

Edith Wharton wrote, "There are two ways of spreading light: to be the candle or the mirror that reflects it." My father is my candle; throughout my life I have and will continue to be the mirror that reflects his light.

Rest in Peace Papa

Yours Truly,

Dr Shivani Agarwal

Contents

Preface ix
Editors xi
Contributors xiii

Chapter 1 Applications of Industrial Internet of Things (IIoT) 1

Sandhya Makkar, Megha Duseja, and Shivani Agarwal

Chapter 2 Biomedical Applications Using IoT 21

Neeraj Kumar Jain, Preeti Mittal, and Rajesh Kumar Saini

Chapter 3 Emerging Technological Advances in Healthcare 45

Sweta Silpa Mohapatra, Ashirwad Kumar Singh, and S. Shiva Koteshwar

Chapter 4 The Internet of Things (IoT) and Contactless Payments: An Empirical Analysis of the Healthcare Industry 61

Pooja Ahuja and Sandhya Makkar

Chapter 5 AI in Healthcare 77

Ghous Bakhsh Narejo

Chapter 6 Security Vulnerabilities in the IoT 93

Neeraj Kumar Jain, Preeti Mittal, and Rajesh Kumar Saini

Chapter 7 Research on IoT Governance, Security, and Privacy Issues of Internet of Things 115

Manish Bhardwaj

Chapter 8 Recent Trends of IoT and Big Data in Research Problem-Solving 135

Pham Thi Viet Huong and Tran Anh Vu

Chapter 9 A Theoretical Context for CSF in Medical Software Next Release 163

Hemlata Sharma, Shushila Madan, Nisheeth Joshi, and Nidhi Mathur

Chapter 10 Robotics and Machine Learning 183

Ghous Bakhsh Narejo

Chapter 11 Detecting Medical Reviews Using Sentiment Analysis 199

Sandhya Makkar, Mayank Singhal, Nimish Gulati, and Shivani Agarwal

Index 217

Preface

The Internet of Things (IoT) opens opportunities for wearable devices, home appliances, and software to share and communicate information on the Internet. In the past, companies could rely on confining themselves to a particular geographical area to conduct their business. Today companies are increasingly becoming location-independent and are finding themselves to be strategically disadvantaged if they are restricted to a particular place. The consequence of advances in information technologies and the changing boundaries of businesses have brought the importance of data and information to the fore. It is information that helps companies realize their objectives and helps managers make appropriate decisions. Given that the shared data contains a large amount of private information, preserving information security on shared data is an important issue that cannot be neglected. This has motivated editors to investigate opportunities to introduce new technology in organizations through research work by practitioners and innovators.

With the same aim, the editors of this book circulated a call for chapters to researchers, entrepreneurs, and academics and received a number of quality chapters. However, due to the scope of book and review reports, we were only able to include 11 chapters in the edited book. These articles exemplify the analysis and exploration of complex models and data sets from various domains. They provide invaluable insights into the studied problems and offer convincing case studies and experimental analysis. We are thankful to the authors for enduring a long wait, and finally the result is before you.

In this book, we begin with general information on the security background of Industrial IoT (IIoT) and continue by examining the information security related applications and challenges that IoT encounters. We also point out research directions that could be the future work toward solutions to the security challenges that IoT encounters. The following few chapters present the applications of artificial intelligence and IoT in various domains like healthcare, telecommunication, and so on.

In another section, an empirical analysis of the healthcare industry with a special focus on the role of IoT in digital or contactless payments was presented. The chapter describes contactless payments, popularly referred to as *cashless payments* or *e-wallets*. The paper presents the advantages of the association of contactless payments with IoT in healthcare systems in Asian countries. The empirical study explores the acceptance of new technology in healthcare clinics and hospitals using the Technology Acceptance Model (TAM). The following chapter discusses the role of Artificial Intelligence in healthcare and how it might impact the world.

The next section of the book is based on security vulnerabilities and privacy challenges. This section discusses four key points: IoT governance, security and privacy issues in IoT, recent trends in IoT and Big Data, and the role of IoT in robotics and machine learning. Finally, some future directions and further research are mentioned.

We offer our sincere thanks to reviewers at home as well as from abroad who have given their valuable time to provide critical reviews for the assigned book.

We are sincerely thankful to Taylor & Francis Group for providing the opportunity for us to edit this book. I am sure that this book will be beneficial to the industry, academicians, and researchers who are working in the domain of human resources, organizational behaviors, artificial intelligence, analytics, and so on.

We are also thankful to our respective institutions, the principals/directors of KIET, Ghazhiabad, India; Lal Bahadur Shastri Institute of Management, Delhi, India; and Phenikaa University, Hanoi, Vietnam, who provided us ample open platform to fulfill our editorial duties honestly. I am sure this learning will be helpful for us in research and academia.

The utmost care has been taken by us in preparing this volume, but nevertheless, critical feedback and suggestions from readers will help us to get better inputs in future endeavors.

Shivani Agarwal
Sandhya Makkar
Duc-Tan Tran

Editors

Shivani Agarwal is an assistant professor at the KIET School of Management at KIET Group of Institutions, Ghaziabad, India. She earned a PhD in Management from the Indian Institute of Technology (IIT, Roorkee). Prior to her current role, she was associated with the Institute of Technology and Science, Ghaziabad, UP, India, HRIT Group of Institutions, Ghaziabad, UP, India; IIT Roorkee, UK, India; and Center for Management Development, Modinagar, UP.

She has attended several short term courses at the Indian Institute of Technology Roorkee, India, IIT Delhi, India; she has also earned NPTEL certificate for Research Writing and Human Resource Management.

She has authored or co-authored more than 10 research articles that are published in journals, books, and conference proceedings. She teaches graduate- and postgraduate-level courses in management.

She earned a PhD in Organization Behaviour (Psychology) from IIT Roorkee, India, in 2015; an MBA from Guru Gobind Singh University, New Delhi, India, in 2009; and a bachelor's degree in Science from Chaudhary Charan Singh University, Meerut, India, in 2007.

She is the Series Editor of "Information Technology, Management and Operations Research Practices," CRC Press, Taylor & Francis Group, USA. She is guest editor with IGI-Global, USA. She can be contacted at jindal.shivani24@gmail.com.

Sandhya Makkar is working as a senior assistant professor in the Department of Operations and Information Systems, Lal Bahadur Shastri Institute of Management, Delhi, India. She has more than 13 years of academic and research experience in Management Science, Supply Chain Optimization, Machine Learning, and Big Data. She remained first class throughout her academic career. She holds university rank during graduation (Mathematics) and post-graduation (Operational Research). She earned a doctoral degree (PhD) in the area of Supply Chain Optimization from the University of Delhi, India (DU, Delhi) and has many research papers and articles to her credit. She is actively involved in research on Supply Chain Optimization, Artificial Intelligence, and other related areas.

She is an area convener of the Logistics and Supply Chain Programme (L&SCM) Programme at Lal Bahadur Shastri Institute of Management. She has authored or co-authored numerous research articles that are published in international journals, books, and conference proceedings. Currently she is editing several books in the area of Management and Artificial Intelligence. She was also appointed as Series Editor of the series "Information Technology, Management and Operations Research Practices," CRC Press, and another series, "Big Data for Industry 4.0: Challenges and Applications," CRC Press, Taylor & Francis Group, USA. Dr Sandhya has also reviewed various research papers in the same area for various conferences and journals. She is a member of the editorial team of American Association for Science and Technology (AASCIT). She has also presented her work at many international conferences.

Duc-Tan Tran is an associate professor and vice-dean of the Faculty of Electrical and Electronic Engineering (FEEE), Phenikaa University, Hanoi, Vietnam. From August 2016 to May 2019, he was an associate professor and vice-dean of the Electronics and Telecommunication Faculty, VNU University of Engineering and Technology. He has published over 150 research papers. His publications received the "Best Paper Award" at the 9th International Conference on Multimedia and Ubiquitous Engineering (MUE-15) and the International Conference on Green and Human Information Technology (ICGHIT-2015). He was the recipient of the award for excellent young researcher from Vietnam National University in 2008, Hanoi, and third prize in the contest "Vietnamese Talents" in 2008. His main research interests include the representation, processing, analysis, and communication of information embedded in signals and datasets. He serves as a technical committee program member, track chair, session chair, and reviewer for many international conferences and journals. He is a TPC co-chair of the 2016 International Conference on Advanced Technologies for Communications (http://atc-conf.org/); the 2018 National Conference on Electronics, Communications and Information Technology (http://rev-conf.org); the 2019 International Conference on Research in Intelligent and Computing in Engineering (http://www.riceconference.in); and the 2019 International Conference on Advanced Technologies for Communications (http://atc-conf.org/).

Contributors

Shivani Agarwal
KIET School of Management
KIET Group of Institutions
Delhi-NCR, India

Pooja Ahuja
Department Research Scholar
Swami Vivekananda University
Sagar, India

Manish Bhardwaj
Computer Science and Engineering
KIET Group of Institutions
Delhi-NCR, India

Megha Duseja
Lal Bahadur Shastri Institute of
    Management
New Delhi, India

Nimish Gulati
Lal Bahadur Shastri Institute of
    Management
Delhi, India

Pham Thi Viet Huong
International School
Vietnam National University
Hanoi, Vietnam

Neeraj Kumar Jain
Department of IT
Institute of Technology and
    Science
Mohan Nagar Ghaziabad, India

Nisheeth Joshi
Banasthali Vidyapith
Jaipur, India

S. Shiva Koteshwar
Research and Business Analytics
Lal Bahadur Shastri Institute of
    Management
New Delhi, India

Shushila Madan
LSR
Delhi University
Delhi, India

Sandhya Makkar
Lal Bahadur Shastri Institute of
    Management
Delhi, India

Nidhi Mathur
IMT-CDL
Ghaziabad, India

Preeti Mittal
Department of Computer
    Applications
IIMT University
Meerut, India

Sweta Silpa Mohapatra
Research and Business Analytics
Lal Bahadur Shastri Institute of
    Management
New Delhi, India

Ghous Bakhsh Narejo
Department of Electronic
    Engineering
NED University of Engineering and
    Technology
Karachi, Pakistan

Dr Rajesh Kumar Saini
Department of Mathematical Studies and Computer Science
Bundelkhand University
Jhansi, India

Hemlata Sharma
IMT-CDL
Ghaziabad, India

Ashirwad Kumar Singh
Research and Business Analytics
Lal Bahadur Shastri Institute of Management
New Delhi, India

Mayank Singhal
Lal Bahadur Shastri Institute of Management
Delhi, India

Tran Anh Vu
School of Electronics and Telecommunications
Hanoi University of Science and Technology
Hanoi, Vietnam

1 Applications of Industrial Internet of Things (IIoT)

Sandhya Makkar, Megha Duseja, and Shivani Agarwal

CONTENTS

1.1 Introduction 2
    1.1.1 Internet of Things..2
    1.1.2 Machine-To-Machine (M2M)..2
    1.1.3 Emergence and Growth of IoT 3
    1.1.4 Setting up an IoT Infrastructure...3
1.2 Cloud System Selection 3
1.3 Platform Selection..5
1.4 IoT Applications in the Industry 5
    1.4.1 Vehicle Telematics...5
    1.4.2 Fleet..6
    1.4.3 Connected Car..7
    1.4.4 Leasing/Ride Sharing/Asset Management..7
    1.4.5 Healthcare 8
    1.4.6 IoT for Patients 9
    1.4.7 Medical Adherence 9
    1.4.8 IoT for Physicians 10
    1.4.9 Blood Banks..10
    1.4.10 Device Monitoring: Defibrillators/Heart Monitors/Pacemakers 11
    1.4.11 IoT for Hospitals 11
    1.4.12 IoT for Health Insurance Companies 11
    1.4.13 Smart Metering 12
    1.4.14 Smart House..12
    1.4.15 Smart City 13
    1.4.16 Renewable Solar Energy 13
    1.4.17 Insurance 14
    1.4.18 Enterprise Use Cases..14
    1.4.19 Using Iot for a Better World 15
1.5 Security, Privacy, and the Internet of Things 15
    1.5.1 Multiple Network 16
    1.5.2 Multiple Types of Services...16
    1.5.3 Scaling Growth 16

1.5.4 Automated Functionality 17
1.5.5 Long Lifecycles 17
1.5.6 Remote Updates 17
1.6 The Future of the Internet of Things 17
1.6.1 IoT Will Be Primary 17
1.6.2 Homes Will Get Smarter 18
1.6.3 Enterprises Will Start Using More IoT 18
Bibliography 18

1.1 INTRODUCTION

1.1.1 INTERNET OF THINGS

The Internet of Things, or as it is informally referred to, *IoT*, is a system of devices, people, and machines all well connected wirelessly through the internet and to each other.

The perception and uses of the term *IoT* will continue to evolve with the changing times as new connected technologies are developed and pave the way for more such applications. The possibilities just keep growing.

The need for more advanced solutions will drive the next IoT innovations in a world where consumer satisfaction is imperative for organizational growth. Every business can leverage the digitally connected ecosystem of IoT devices to have that technological edge in this furiously competitive world we live in. Not only will IoT enhance efficiency but it can also provide valuable insights into issues related to streamline workflows, predict necessary maintenance, analyze usage patterns, automate manufacturing, and much more.

1.1.2 MACHINE-TO-MACHINE (M2M)

M2M technologies represent closed, point-to-point communications between machines or between machine and management systems, without the need for human intervention. M2M devices, enabling bidirectional remote monitoring and transfer of data, consist of a sensor or a radio frequency identification (RFID) tag and a communication module.

M2M devices, as the industrial precursors to the IoT, can include items ranging from in-house/in-office machinery, such as printers or scanners, to manufacturing equipment, including heavy machinery.

But do not assume that the IoT will replace M2M. Predictions show that cumulative M2M connections will grow from 995 million in 2014 to a projected 2.7 billion connections by 2018.

M2M uses include telemetry, traffic control, security, tracking and tracing, machinery maintenance and control, metering, and manufacturing and facility management, as well as a multitude of other applications.

M2M/IoT will represent the next phase of the internet revolution, connecting about ten times more devices on the internet in a couple of years. So, it is best to be aware and start embracing it from its infancy and gain the first mover advantage.

Industrial Internet of Things

1.1.3 Emergence and Growth of IoT

The IoT establishes an end-to-end ecosystem, including technologies, processes, and concepts employed across all connectivity use cases.

In principle, modern information and communication technologies play a vital role in IoT. As IoT is nothing but an exhaustive web of recognizable physical things, this will involve connecting not only humans but objects as well. These objects come with small processors and sensors like temperature controllers and audio or video output devices that are used to supervise the surrounding objects. This helps to make the connected objects more responsive to the changes in the surroundings and thus to analyze the appropriate information or service to be provided, which makes our objects "smart".

The foundation of IoT is automatic identification through RFID. Sensors and actuators help by enhancing the applicability by encapsulating the real-time effects of actions. This leads to a huge opportunity for new services and applications in many areas. For instance, consumer goods may have a large repository of data which will help in providing a faster and more specific personalized service later on. Not only will this result in new business models and product innovations but it will also create a new line of products and services for more narrowed-down and specific customer needs, thus having a huge influence on individuals, markets, society, and enterprises.

The term *Internet of Things* was coined by Kevin Ashton while presenting RFID technology connected to the internet to easily identify products at Proctor and Gamble. The notion evolved and came to be more widely used in real life during the 2000s, especially after 2008, when the IPSO Alliance was formed to encourage the use of networked objects/things. Table 1.1 gives a summary of the progress of IoT according to Forbes.

1.1.4 Setting up an IoT Infrastructure

Before setting up an IoT infrastructure, the project manager should thoroughly investigate the viability of the project and whether the IoT solution will lower the costs of the business. The manager should also thoroughly investigate the factors affecting the setup.

1.2 CLOUD SYSTEM SELECTION

Often, the large number of devices that are typically deployed for an IoT application necessitates the use of a commercial cloud service that provides the performance and high-availability capability required by the application.

This requires customers to carefully ascertain whether the cloud provider has the right tools, cost models, and support for their needs. The right questions are not always clear since some aspect of their service or cost model may be appropriate until a certain size threshold is reached. This "scaling challenge" is often one of the most difficult areas of assessing what is possible.

TABLE 1.1
Historical Journey of IoT

Year	Event
1949	Norman Joseph Woodland invents the barcode, which is patented in 1952 and is first used by supermarkets 20 years later.
1955	Edward O. Thorp creates the world's first wearable computer, whose only function is to predict roulette wheels.
1967	Hubert Upton develops an analog wearable computer.
1969	The first message is sent over ARPANET.
1973	Invented by Mario Cardullo, the RFID tag receives its first patent.
1977	C.C. Colins creates a product for the visually disabled, a wearable that transforms an image into a tactile grid on a vest.
Early 1980s	Carnegie-Mellon Computer Science Department researchers install micro-switches in the Coke vending machine and connect them to a computer to make supervision easier.
1990	John Romkey comes up with a toaster connected to the internet.
1990	Olivetti invents a badge system to locate people.
1993	Thad Starner from MIT uses a heads-up display as a wearable.
1993	Steven Feiner, Blair MacIntyre, and Dorée Seligmann invent KARMA—Knowledge-based Augmented Reality for Maintenance Assistance.
1995	GPS Satellite Network (version 1) is completed.
1999	An Automatic Identification Center is set up at MIT where objects are linked to the internet through the RFID tag.
1999	The term *Internet of Things* is first used by Kevin Ashton of MIT.
2000	LG designs the first "smart fridge".
2003	Publications like *The Guardian*, *Scientific American*, and *Boston Globe* mention the Internet of Things in their articles.
2004	US Defence uses IoT in their Savi Program, and Walmart starts using IoT in commercial retail.
2005	The Interaction Design Institute Ivrea (IDII) invents a small microcontroller to help their students in making their projects.
2007	The United Nations' International Telecommunications Union publishes a report on Internet of Things, saying "from anytime, anyplace connectivity for anyone, we will now have connectivity for anything. Connections will multiply and create an entirely new dynamic network of networks—an Internet of Things".

Consider the following:

- **Does the cloud provider have a cost model that matches what the IoT application can bear?**

  Can the cloud provider assist you in simulating the costs for your application transport, storage, and analysis needs? If the costs do not meet expectations, the return on investment for the application could fall short and lead to an unsuccessful deployment.

- **Does the cloud provider offer data centers in the countries where the devices are deployed?**

Industrial Internet of Things

In some countries, there are regulations that require that data must not cross national boundaries, and the presence of a local data center may be critical to operating within the regulations of that country. Indeed, an absence of a cloud data center in a vital market may preclude the selection of that provider.

- **Does the cloud provider have the tools to support high-availability deployments? Do your software engineers and operations personnel have the expertise to develop and maintain cloud solutions for your IoT application?**

Sometimes, the selection of a cloud provider is guided by the available personnel within your company who have experience with that provider. It may be necessary, however, to hire additional resources or use an IoT platform vendor who can guide you to the best possible solution.

1.3 PLATFORM SELECTION

Many companies attempt to provide a platform for IoT solutions. This appears to be an area where it is possible to find hundreds of companies purporting to provide "IoT platforms".

In this noisy environment, it is difficult to assess what the capabilities and features of the platform are, let alone how well they would fit for the requirements of your specific IoT deployment.

Given the large variety of possible IoT applications in many different types of markets and businesses, and the large number of available platforms, it is tough to determine the best one for your needs. Yet, it is important to make the best selection as early as possible, since the wrong selection at the early phase of any IoT application deployment could significantly impact and delay the project.

1.4 IOT APPLICATIONS IN THE INDUSTRY

1.4.1 Vehicle Telematics

Telematics is a mechanism for tracking a vehicle. By simply amalgamating a GPS system with on-board diagnostics it is possible to record and find out exactly where a car is at that particular moment and its speed, while, at the same time, to track how the car is functioning internally.

By enabling connectivity with the help of a SIM, telematics can be used to transmit data back and forth between a vehicle and a central management system. This technology has been used by Formula One teams for years to track their opponents during a race, all through just a sensor in the cars and a trackside wireless network. A lot of dedicated research has been done on this topic as well as on other applications of IoT by Syed Zaeem Hosain, chief technology officer at Aeris, and this is presented in his book titled *IoT in the Business*, a few examples of which I will be citing here.

Speed, acceleration, average time and distance driven, and all the other relevant metrics can be easily recorded with the help of a small, reliable onboard unit (OBU).

These metrics can then be used to analyze behavior much more accurately than demographic profiling and at the same time to target marketing programs and gain further knowledge into consumer behavior.

For insurers, the benefits of telematics are the following:

- Fewer fraud cases because of more accurate tracing as well as better data when assessing claims
- A more responsible and accountable driving behavior, which is a result of drivers being motivated to repeatedly practice and inculcate better driving habits, thus leading to lesser accidents as well
- Stronger and more long-lasting customer relationships, with constant and constructive interactions (e.g., communications after an accident, updates on any problems) instead of just the usual day-to-day paperwork.

Let us delve more into how this benefits the automotive industry and the various components involved in the whole process.

1.4.2 FLEET

In a world where even a minute's latency in solving a transportation issue can lead to a very costly mistake, real-time intelligence provision is the need of the hour and is here to stay for good! A connected data transport solution can be combined with IoT analytics to reduce time-to-market processes, resolve troubleshooting issues, and bring down the total cost of ownership. It is only with real-time business intelligence data that retailers and manufacturers can acquire a comprehensive view of their transportation ecosystem.

The global trucking industry is currently undergoing extensive change. Very soon, older trucks will be replaced with "smart trucks" using IoT systems with 5G and VSAT technologies to communicate vital information, ensuring better supervision.

Moreover, commercial fleet solutions are turning out to be trickier than was previously thought as new needs like transregional connectivity and goal-oriented solutions are emerging in the industry.

Long-haul fleet management providers require onboard computing as well as fleet communications in order to deliver better results and to stay updated. Onboard solutions require highly reliable, real-time, always-on cellular network connectivity, which might require multiple carriers to meet the full-coverage needs of fleet customers wherever they reside and drive.

The market for global fleet management solutions continues to expand quickly at a very fast pace. Transparency Market Research concluded that the fleet management solutions sector will rise in value from US$12.5 billion in 2015 to an expected US$92 billion by 2025, with a compound annual growth rate (CAGR) of 22.6% for the period between 2017 and 2025.

In North and Latin America, according to Berg Insight, connected fleet management will grow from 5.8 million rolling units in operation to more than 12.7 million by 2020. Additionally, Berg Insights see similar European growth in the sector, expanding from 5.3 million units in 2015 to an estimated 10.6 million by 2020.

As the commercial fleet sector becomes even more competitive, fleet owners and operators are scouting for more trustworthy relationships, ones that provide cooperative pricing and better performance, in order to make the most of the disruptive opportunity made possible by IoT.

1.4.3 CONNECTED CAR

According to recent surveys, 75% of all new cars shipped globally by 2024 will be equipped with wireless connection. Nearly 20 years after General Motors shipped its first connected Cadillacs, car companies have learned that it is not enough to only build a telecommunications module into the vehicle. The entire system supporting that connection must be properly designed and maintained for the connected car proposition to be viable, and all this should be done in such a manner that is convenient as well as aesthetic.

While the requirements of instantaneous action, such as accident avoidance, means that the cars must process sensor data extremely rapidly——local to the vehicle——other capabilities will be enabled by faster cellular technologies, such as 5G. Hence, a proper integration with cellular companies is also of prime importance.

For example, updates for general traffic conditions beyond the range of vehicle-to-vehicle (V2V) radio technology, as well as dynamic updates for road changes (repair work, hazards), can be enabled by the advent of faster cellular technologies. Today, connected cars are utilizing IoT for different engine diagnostics, global navigation satellite systems (GNSS) location data, and suggested alternative routes in order to avoid traffic. Modern connected car systems are expected to be around for a long time and, increasingly, will be asked to support autonomous driving and safety applications, such as collision avoidance and congestion avoidance.

For instance, a reactive system like OnStar's original offering is no longer enough. Today, 80% of connected cars are using technology that is more than a decade old and hardly suited to automakers' rapidly changing needs, let alone those of the customer. This is even more true for developing countries like India where the majority of the population falls into the middle-income group; however, this can be overcome easily if correct steps are taken by the government and automobile brands offer good value proposition.

Wireless connectivity, today and into the future, will be expected to maintain continuous connections, with increasing demands on the communication of vehicle data, software updates to protect vehicle data, and greater insights leading to a safer and more efficient driving experience.

All of which can be of great help for taxi and cab providers like Uber and Ola.

The future of fully autonomous driving is interwoven with the communication technologies in development to make this a practical reality.

1.4.4 LEASING/RIDE SHARING/ASSET MANAGEMENT

In many parts of the world, transportation is a huge challenge. Traffic, costs, weather, and vehicle availability all come into play. To further alleviate these issues, ride-sharing companies are rapidly expanding globally, trying to fill the need to move people in urban areas.

In fact, revenue from the global ride-sharing sector is approaching US$57 billion and is expected to grow at an annual rate of 16.5%, leading to a market of US$106 billion by the year 2022.

In many countries, drivers want to lease vehicles so they can create their own ride-sharing business driving for a specific brand.

This creates a significant risk of loss of the leased asset due to theft, lack of payment, or hijacking.

Additional loss can come from misuse of the vehicle.

In order to protect their investments, ride-sharing companies need to track all their vehicles, many of which are leased to individual drivers. Companies need data on driver performance and access to vehicle metrics regarding whether the leased vehicle is being used for another service or whether the driver is paying leasing fees.

In such cases, ride-sharing companies also need the ability to remotely disable the vehicle before harm can be done to a company or its reputation.

The key to a smooth business is reliable monitoring and tracking connectivity so that data is collected in a timely manner and in such a way that it is accurate and precise.

Poor quality IoT devices and slow response times to problem-management scenarios are situations that also need attention.

Today, ride-sharing companies can install a tracking device in their vehicles, which ensures constant monitoring of vehicle location; insights into driver performance metrics and their reliability and safety; vehicle metrics; an auto-immobilize functionality if the driver is late in paying leasing charges or if the vehicle is reported stolen or tampered with; and the possibility of stopping the driver using the vehicle for unauthorized purposes. This comprehensive IoT asset management solution allows ride-sharing enterprises to retrieve any vehicle operating outside of company guidelines and to protect its investment in a costly asset.

1.4.5 Healthcare

It has been estimated that around 40% of the global economic impact of the IoT revolution will occur mainly in healthcare, as this sector is the one which requires the most surveillance and sensors. IoT-driven companies can gain a huge competitive edge in that sector—specifically in areas such as user experience, operational costs and efficiencies, and global expansion.

With enormous potential to help patients as well as doctors, IoT-enabled devices are certainly the next big thing in healthcare and are here to stay for a very long time.

Along with making patients feel safe and protected, these devices will also lead to greater patient satisfaction since they make interactions with doctors easier and more convenient.

The devices also help to shorten hospital stay length by providing real-time recovery assistance.

Industrial Internet of Things

1.4.6 IoT for Patients

Estimates show that more than 200 million people in the European Union and the United States suffer from one or several diseases that may benefit from some type of home monitoring.

New IoT technologies are revolutionizing the way healthcare services are provided, enabling patients to stay at home and receive the same service that would be provided at a hospital. Healthcare companies are extending their in-home services, delivering an easier life to those living independently, specifically tailored to serve senior citizens and disabled people worldwide. With that in mind, more and more state and federal healthcare agencies are promoting at-home care solutions as an essential way to enhance efficiencies and reduce costs.

Let us look at such a company named Simply Home, which designs and installs wireless/Wi-Fi technology products and similar healthcare-focused services. The company deploys a cost-effective IoT cellular solution that provides connectivity for its services, regardless of the patient's location. Its systems proactively alert patients and caregivers to alterations in day-to-day behavior by communicating with multiple sensors to observe a patient's everyday activities.

Text, email, or phone notifications can be triggered by a single activity, or many such activities, or by inactivity. Elements such as motion sensors, door/window contacts, and bed pressure pads alert caretakers to falls, wandering, or changes in sleep patterns.

The IoT-enabled Simply Home system helps residents stay carefree with environmental controls that switch on beds, lights, TVs, doors, and more via tablet or voice activation.

IoT is nothing short of a major milestone in elderly healthcare, mainly because of its feature of keeping a real-time track of the patient's health. It is a boon for elderly people living on their own or even with their families, as even a slightest change in the routine activities of the patient alerts both their loved ones and doctors.

1.4.7 Medical Adherence

Medical adherence is a topic which is mostly ignored and taken for granted, but it can be a leading cause for lower immunity and deteriorating health.

According to a recent research by the World Health Organization (WHO), the benefits of medications used to fight disease are not fully realized only because close to 50% of patients do not adhere to medicinal intake guidelines and, in most cases, fail to adhere to the timings prescribed by the doctor. The causes for not taking medicines on a regularly needed basis are plentiful, ranging from lack of funds, suboptimal healthcare literacy, or communication/language barriers to a busy lifestyle or just plain forgetfulness or a casual attitude.

Wisepill, a company which has assisted doctors, nurses, and researchers for many years to provide medical adherence management solutions, have come up with the Wisepill dispenser, an easy-to-use medicine box that uses cellular and IoT technologies to provide daily medical adherence assistance. The pillbox, designed to work in diverse environments, has a rechargeable, longer-life battery, which allows the device to be used for extended periods without the need for an external power source. Patients in developing countries or in hard-to-reach rural areas cannot travel easily to far-off clinics. Additionally, many places have a severe shortage of medical professionals, like rural areas in India. With IoT connectivity, Wisepill enables clients, pharmaceutical businesses, doctors, and healthcare organizations around the world to improve medication adherence management. The technology used in the device facilitates researchers to retrieve data on a daily basis, so that the data does not get lost even if the device goes missing, and ensures that feedback is processed promptly.

The combination of an experienced IoT solution provider and Wisepill provides patients with the peace of mind of knowing that if they miss taking their medications, there will be a reminder to maintain their medicinal intake schedule.

1.4.8 IoT FOR PHYSICIANS

Physicians can track patients' recovery and speed of recovery as well as real-time health updates very effectively with help of IoT-enabled wearables and other home monitoring equipment; this will also help the physician to track whether the patient is taking their prescriptions properly and on time, and this will also alert the physician in case any special attention is required.

1.4.9 BLOOD BANKS

Blood units form a pivotal area of healthcare. Yet often, blood units go to waste due to the inability to store them under appropriate conditions, mainly because of the required low temperature to store it. The principal goal of an IoT technology–driven blood-bank management program is to optimize the effectiveness of a blood bank.

A successful program involves increasing awareness about best practices; reducing the likelihood of blood samples becoming unusable; reducing blood loss; enhancing blood availability; regularly informing physicians; as well as standardizing operations through workflows.

However, it should also be considered how implementable such a program is.

In India, the business case for an IoT-enabled blood-bank monitoring solution rests on the following goals:

- Monitoring of blood-bank refrigerators on a 24 × 7 basis and storing the relevant data
- Alerting temperature variance outside a set range through the help of sensors
- Using a Transparent Monitoring Network (single pane of glass)
- Reducing paperwork.

The IoT-based blood-bank improvement program includes a measurement of how effectively the program reaches its goals and also indicates cause-based results.

This, plus additional functionality, provides the insights needed to initiate and preserve blood-bank management that will save many lives.

1.4.10 Device Monitoring: Defibrillators/Heart Monitors/Pacemakers

Today, IoT-connected medical devices can monitor and analyze data coming from the patient in real time.

Should an event occur, IoT-enabled defibrillators, for example, provide verbal and on-screen instructions for delivering chest compressions. Some advanced defibrillators even can deliver an electrical shock to a patient's heart. Heart monitors send alerts to both wearer and doctor in case of irregularities.

In all these cases, device connectivity, with real-time data, is literally a matter of life and death and, in other words, is a life saver for many.

With an IoT-enabled solution for healthcare devices, real-time data, along with alerts and reports, can save lives while at the same time keeping people connected to each other.

1.4.11 IoT for Hospitals

Other than supervising patients' health, IoT devices have a plethora of other applications in healthcare. IoT devices which have sensors can be used for tracing exactly where they are and can even be used to track medical equipment like wheelchairs, defibrillators, nebulizers, oxygen pumps, and other monitoring equipment. Medical staff like doctors and nurses can also be traced.

Secondly, the spread of infection is one of the main concerns for patients in hospitals. Devices with IoT can also be a major factor in preventing patients from easily catching an infection.

However, there is also a huge scope for personnel management in helping to keep a real-time record of nurses, doctors, and patients, thus creating a seamless ecosystem of faultless executions and operations.

1.4.12 IoT for Health Insurance Companies

IoT provides a world of potential for health insurance companies as they can make a good use of the data encapsulated through health monitoring IoT devices for underwriting and make claims more accurately. Moreover, it will also help the companies detect whether claims are genuine or false and hence identify prospects for underwriting and finding loopholes, if any.

IoT provides a lot of assistance in areas like underwriting, pricing, risk assessment, and claims handling. This not only gives customers much more clarity but also allows them to see the bigger picture.

Let us look at such real-time use cases of IoT:

In Los Angeles, everyday congestion on roads is a very common sight and this has been the case for decades. Therefore, information gleaned using magnetic road sensors and hundreds of cameras helps in keeping around 450 traffic signals congestion free. The system, which is worth around $400 million and was instigated in 2013, also has many other benefits like accelerating travel speed around Los Angeles by 16%, thus lessening delays at many crucial signals by around 12%.

In San Francisco, SFpark leverages wireless detectors to discover parking-space occupancy in metered spaces. Installed in around 8,200 on-street places in the city of San Francisco, these wireless detectors trace parking-space vacancy in real time. Two years after its launch in 2013, SFpark, San Francisco came up with a detailed report clearly showing that their initiative had lessened greenhouse-gas emissions by almost 25%. Not only this, they also reported a significant reduction in traffic volume as well as driving time. It also made it very simple to make payments for parking and led to a considerable decrease in loss due to broken parking meters, which in turn led to a rise of around $1.9 million in parking revenue for San Francisco.

Major cities like London have started testing a smart parking report that enables drivers to track parking spaces in real time, considerably reducing the time spent searching for open spots to park their cars. Such steps help in many ways, like significantly reducing traffic, reducing fuel consumption, and lowering harmful carbon emissions. In 2011, a company called Autolib started an electric car sharing campaign in Paris that has grown to over 3,000 vehicles in a span of a few months. The cars connected through GPS can be traced live to get real-time updates about their location, and the car's dashboard can be used to reserve parking spaces in advance, which not only saves time but also lessens the amount of waste and the carbon footprint associated with parking spot searches.

The city of Copenhagen uses sensors to track the city's bike traffic in real time with the help of GPS, thus providing useful insights on how to upgrade the bike routes in the city. This is vital as more than 40% of commuters use a bike every day.

In an attempt to save water, the drought-plagued town of Fountain View in California implemented programs like FlexNetcommunication system, iPERL residential and OMNI commercial meters to decrease water usage by almost 23% (Sensus.com, 2017).

1.4.13 Smart Metering

Smart metering is among the primary steps to developing a detailed city-wide smart grid system that addresses challenges surrounding not only energy consumption but also water usage. This is possible through the help of real-time data tracking of electricity and gas usage, which is how smart meters enable utility providers to master energy distribution and at the same time also making sure consumers take smarter decisions regarding their energy consumption. Through the help of IoT solutions specially made for smart metering, customer service can also be improved substantially with the help of the insights provided by these IoT solutions, thus creating a win-win situation from every point of view.

Whereas traditional meters can only track total consumption, smart meters not only track total consumption but also provide information about the when and how of the consumption. This is exactly why power companies are now using smart meters to monitor consumer usage and price the energy according to the usage by dividing the consumption into different quotas.

1.4.14 Smart House

To see how IoT applications can cater to the needs of a house deserves a separate mention as there a plenty of products the IoT has to offer in this domain. Among the best applications are air conditioners, speakers, smart thermostats, pet feeders, and many such thoughtful innovations specially innovated to enhance and manage the day-to-day operations in the home. It is thus currently one of the most popular and most-used areas of IoT and is also the most promising, considering its cost effectiveness as well as its efficacy.

1.4.15 Smart City

Practically speaking, every aspect of city operation can be made smarter and more efficient as well as optimum when it is connected through IoT—from hidden roadway devices to parking management to biodegradable waste management. This shows there is a strong future for IoT providers to deliver a plethora of solutions in order to cater to the requirements of efficiency and lower operating costs.

We can all understand the tendency of humans to relocate from one city to another or from a rural area to an urban, more developed city to seek a better life and better employment opportunities as well as an improved lifestyle for themselves and future generations.

However, we often tend to ignore the drawbacks this phenomenon brings, which include the need for better infrastructure as well as the deterioration of the quality of life the existing residents of large, ever-growing cities.

A "smart city" may sound futuristic, but at its heart, the idea is quite simple and traditional—smart cities bring together current and new technologies, infrastructure, and government to benefit people's quality of life.

There are enormous opportunities for IoT in cities, but most plans are at the development stage with application broadly in the domains of smart parking, smart roads, smart lighting, smart waste management, and smart firefighting. In less than 5 years maximum, these words will soon be a reality, thus making the planet eco-friendlier and less chaotic.

1.4.16 Renewable Solar Energy

According to the World Energy Outlook, around 1.2 billion people, roughly 16% of the global population, do not have access to electricity; about 95% of those people live in sub-Saharan Africa and developing Asia, and 80% live in rural areas.

Hence, the need of the hour is to provide affordable, clean, and high-quality energy. For instance, BBOXX, a UK-based solar energy provider using IoT technologies, has developed solutions to provide energy to the less privileged in developing countries. By working with a carrier-agnostic and technology-agnostic partner, BBOXX installed a global subscriber identity module (SIM) at the point of manufacture, leading to both reduced supply costs and faster implementation.

Following suit is an Israeli company called SolarEdge with headquarters in the United States, which provides solar solutions for homes and businesses. Company products include power optimizers, solar inverters (DC to AC inversion), and cloud-based monitoring solutions.

With a presence in 13 countries, all with somewhat different connectivity options and providers, secure and reliable connectivity was integral to the company's success. SolarEdge chose a carrier-agnostic and technology-agnostic IoT partner for deploying cellular connectivity, as well as for corresponding management solutions. Dealing with a seasoned, professional connectivity solution provider,

SolarEdge was able to secure instant global connectivity and management oversight, advanced revenue-grade metering, and robust, real-time troubleshooting. Visibility into devices improved, operational efficiency rose, customer service radically upgraded, and inventory and loss management became transparent and easy to administer.

1.4.17 Insurance

A while ago, insurance companies used vehicle telematics for risk assessment, vehicle performance, reports, mobile apps, and application programming interfaces (APIs). However, we can now see a shift in this philosophy, as with sensors and devices absorbing data at an unprecedented rate, insurance now also covers accident reconstruction, false claims identification, overall claims management, driver coaching, alerts and notifications, actuarial support, vehicle immobilization, asset protection, usage-based insurance (UBI), and a lot more.

A platform-based, highly scalable solution enables insurers to offer value-added services that simplify the insurance process and radically improve customer satisfaction.

- *Claims management*: Quickly process claims and identify possible fraud. Reduce fraud and claims while speeding up the entire claims process.
- *Customer management*: Maximize customer value through targeted upsell, cross-sell opportunities. Attract more low-risk customers.
- *Renewals management*: Identify customers with high propensity to lapse for targeted collection. Increase customer retention.
- *Sales force management*: Identify agents with high potential.
- *Pricing and risk management*: Conduct risk-based pricing for better profitability. Gain a higher percentage of low-risk drivers. Reduce underwriting costs. Provide customer premium savings.

In a nutshell, IoT can have a huge impact on services by providing insights for areas like risk assessment, loss control, driver behavior, product pricing, and much more.

1.4.18 Enterprise Use Cases

Ride sharing: Uber is fast becoming synonymous with IoT ride-sharing services, regardless of the location or the mode of on-road transportation. Combining cars, shared carpools, decentralized scooters and bikes, car rentals, and even public transportation, Uber tries to find the most efficient way to get you from here to there. Leveraging smart phones as a basis for its business, Uber connects drivers with those needing conveyance. By relating passenger locations with its drivers, Uber can route services optimally to maximize results. And it doesn't end with driver services. Uber continues to invest in self-driving technologies, acknowledging the near-future direction of vehicles. Additionally, Uber uses other modes of transportation for that "last mile" of transit. Bikes, scooters, and Uber Pools all can alleviate traffic congestion while providing mobility services to its clientele.

In fact, Uber is continuing to invest in new people-moving technology, including dockless scooters and bikes (covering over 70 markets with more than 35,000 scooters in service nationally). And within some cities, Uber has even partnered with public transit, selling tickets via its own app. For inventive companies such as Uber, the IoT is their highway to success.

Solar energy: BBOXX offers a lot of solutions like planning, curating, distributing innovative plug-and play, off-grid solar-powered solutions for energy access across Africa and other such countries. Because of the importance of sustainable energy, BBOXX aims to provide 20 million people with access to electricity by the year 2020. Its main product range incorporates a variety of solar-powered battery boxes which help users to charge small appliances, such as lights, mobile phones, refrigerators, or computers. BBOXX has more than 80,000 systems deployed so far across the world.

1.4.19 USING IOT FOR A BETTER WORLD

SweetSense is an enterprise which has partnered with governments and non-governmental organizations (NGOs) to put IoT sensors on water pumps across rural Africa.

According to a report from Rwanda, only 56% of water pumps there were functioning regularly.

However, after employing the SweetSense technology to track the pumps' function through cellular IoT systems and analytics, the water pumps were able to be repaired more quickly, and 91% of the pumps could be kept working on a regular basis.

Another organization, Hello Tractor, helps sub-Saharan African farmers with food production. The company works closely with its partners to create an entire ecosystem, with a sharing platform for income-generating products (tractor leasing) and affordable service offerings. This has enabled more farmers to receive the services or equipment they need to succeed. With more than 500 tractors in operation, another crucial step to Hello Tractor's success was developing a pay-as-you-go plan that farmers could afford and one that the banks and insurers could accept.

The process deployed by Hello Tractor has produced greater efficiencies, higher crop yields, and a proven business model that can be implemented around the globe. Armed with this greater access to business-critical data, Hello Tractor now has plans to expand projects into Bangladesh and South Africa.

1.5 SECURITY, PRIVACY, AND THE INTERNET OF THINGS

During her address to the audience at the Consumer Electronics Show in January 2015, former US Federal Trade Commission chairperson Edith Ramirez said, "Any device that is connected to the internet is at risk of being hijacked."

This is even more important in today's scenario as we are more dependent upon our phones and devices than ever before; hence our data safety becomes our responsibility.

This aspect should be given more importance today because data is highly prone to misuse, especially our personal information like credit card information, addresses, pin codes, and so on.

The need for a reactive cyber-security infrastructure is the need of the hour as many organizations continue to suffer billions of losses due simply to the lack of a responsive body which can take prompt action against the offenders.

In a world where ethical hacking is full-fledged profession, needless to say, hacking is also becoming a serious threat as many such criminal activities continue to occur on a daily basis, creating many issues for enterprises of all types. What is worse, there is no judicial body to approach to address these issues.

This ultimately results in the blame falling on IoT enabled businesses for poorly protecting data and the invaluable information derived from them.

This has led to cases in the automotive and healthcare industries where the customer information has been misused, leading to negative word of mouth for big brands.

This is why the concept of "Security by Design" is the need of the hour.

Most of the evident loopholes in security can be catered to the moment they are recognized. However, it is imperative to realize that risks can never be completely eliminated, and there is no sure-fire solution for every application.

Thus, it becomes a prerequisite to determine the state of the security taken, and further to assess whether the security is suitable for the kind of data involved; the earlier this is done, the better will be the execution. Hence, it should not be taken for granted.

1.5.1 Multiple Network

Certain IoT devices make use of many transport networks instead of one, in order to avoid latency or because of improved technology, leading to issues of security in any one of the transport networks. This means that different modes have different types of risks; for example, Wi-Fi may have different risks associated with it when compared to the traditional LAN network.

1.5.2 Multiple Types of Services

Instead of using different networks, different types of networks can be used together in a single device, which might lead to a different set of authorizations required for facilitating access to the devices. Hence, the focus should be more on reducing the risk of misuse or leaking of useful data.

1.5.3 Scaling Growth

The explosive growth in IoT deployments means that there will be around a billion potential devices within the span of 5–10 years.

Thus, it has become imperative to have proper checks and balances in place to ensure safety and address the security issues since problems could get magnified if they arise at a later stage; this would mean replacing all those billion devices, which would be extremely difficult, if not impossible.

Industrial Internet of Things

1.5.4 Automated Functionality

The majority of IoT applications work on the simple principle of automation programs, which process the data received and perform the desired actions based on the command. Any loopholes in the transmitted data could compromise the data, thereby cascading into a major problem.

Even augmentation of the devices due to a program error could result in excessive transmission through some devices, thereby burdening those processors with incoming data, resulting in slow responsiveness of the system.

1.5.5 Long Lifecycles

IoT devices are particularly deployed for longer periods of time in industrial applications, providing continuous operations throughout their time in use, unlike other devices with shorter life spans.

The prime reason for this is that IoT devices are electrically powered rather than using batteries, which shut down once the battery gets exhausted; even IoT devices with compromised security stay operational over long durations.

1.5.6 Remote Updates

Over-the-air notifications play an important role in the planning and designing process for device updates, not just updating the application features but also updating the security implementations within the devices; this is critical for detecting data breaches, and thereby the ability to reprogram the functionality remotely is critical since the devices may sometimes be in inaccessible locations.

1.6 THE FUTURE OF THE INTERNET OF THINGS

It is quite evident that the IoT has had an irreversible and irreplaceable impact on industry and is here to stay.

It will prove to be even more useful as it is a boon for both consumers and enterprises, paving the way to exploring more opportunities that can help business to grow by easily collecting and interpreting data, and coming up with new products accordingly.

Not only will this result in lower operational costs, but also a better brand reputation by ensuring better feedback from the customer.

1.6.1 IoT Will Be Primary

Since we live in a world surrounded by different products and services, all driven by technology, innovation, and connectivity, it would not be wrong to say that majority of successful companies will be successful only by keeping "IoT First"; that is, products and services will be curated in such a manner so as to accommodate IoT, while at the same time keeping it as a priority. Moreover, projects and targets would also be designed in such a manner that IoT will be the leading factor for consideration.

1.6.2 Homes Will Get Smarter

As the quantum of volume increases, the price of sensors, processors, and networking devices goes down. This results in increasing awareness on the part of customers about products like the concept of smart home since they customers can now afford them, and no one does this better than IoT, which provides more reliable services through security implementations.

It allows customers to check the status of their homes from distance through enabling security system technologies to check, via security cameras, whether doors are closed and to provide information about break-in conditions, thereby providing theft protection.

It also helps in understanding energy usage patterns and energy savings as it allows customers to automate lighting systems, control temperature, and manage home irrigation systems remotely. Home automation systems identify through sensors which rooms are occupied and adjust heating, ventilation and, air-conditioning (HVAC) systems and lighting to ensure energy conversation. Sensors can also measure the moisture saturation level of the ground and enable sprinklers to maintain outdoor temperatures.

The result is increased safety, convenience, and freedom from daily mundane decisions, enhancing the quality of life.

1.6.3 Enterprises Will Start Using More IoT

Currently, many think of IoT primarily in terms of consumer devices and applications, but industry growth shows that enterprises will be spending far more on IoT than consumers.

McKinsey has predicted that of the $12 trillion economic value produced by IoT during the year 2025, the more than 70% of it will be derived from business-to-business (B2B) deployments. Furthermore, the International Data Corporation (IDC) strongly agrees that by 2021, more than 70% of the top 2,000 global corporations will be investing in connectivity management solutions. Within the same time frame, IDC also expects that 75% of enterprises with a positive IoT return on investment (ROI) will use tactical analytics applications to reduce operating costs, while about 25% of companies that invest in decision architecture will increase their revenue share. Just about any way you view it, IoT can make a significant impact on the bottom line.

BIBLIOGRAPHY

Zeinab Kamal Aldein Mohammed, Sayed Ali Ahmed Elmustafa. (2017). Internet of Things applications, challenges and related future technologies. *World Scientific News*, 67(2) 126–148.

Shahid Raza. (2013). Lightweight security solutions for the Internet of Things. Mälardalen University Press Dissertations. http://www.diva-portal.org/smash/get/diva2:619066/FULLTEXT02.

G.O. Odulaja. (2015). Security issues in internet of the things. *Computing, Information Systems, Development Informatics and Allied Research Journal*, 6(1), 33–40.

Jaydip Sen. Security and privacy issues in cloud computing. Innovation Labs, Tata Consultancy Services Ltd., Kolkata, India. https://arxiv.org/ftp/arxiv/papers/1303/1303.4814.pdf.

Internet of things (IoT) (2016, May 3). http://main.cadit.com.sg/iot-centre-2/internet-of-things/.
https://www.wisepill.com/.
https://www.thingworx.com/ecosystem/markets/smart-connected-systems/smart-cities/.
https://www.simply-home.com/medication-dispenser.
https://www.forbes.com/sites/gilpress/2014/06/18/a-very-short-history-of-the-internet-of-things/.
https://ieeexplore.ieee.org/document/6851114.
http://www.connected-living.org/content/4-information/4-downloads/4-studien/20-mind-commerce-smart-homes-companies-and-solutions-2014/mind-commerce_smart-homes-companies-and-solutions-2014_glo_12-2014.pdf.
http://www.7wdata.be/article-general/how-big-data-and-internet-of-things-builds-smart-cities/.
http://conapptel.org.mx/cincog/aeris_iot_2016_ebook.pdf\.
Gabriele Lobaccaro, Salvatore Carlucci, Erica Löfström. (2016). A review of systems and technologies for smart homes and smart grids. www.mdpi.com/journal/energies.
Eiman Al. Nuaimi et al. (2015). Applications of big data to smart cities. *Journal of Internet Services and Applications*, 6(35), https://doi.org/10.1186/s13174-015-0041-5.
Deshmukh Anup. (2018, November 26). What is Internet of Things? https://dspworks.in/what-is-internet-of-things-iot/.
Amrita Sajja, D.K. Kharde, Chandana Pandey. (2016). A survey on efficient way to live: Smart home – It's an internet of things. *ISAR – International Journal of Electronics and Communication Ethics*, 1(1), 7–12.
Amine Rghioui, Abedlmajid Oumnad. (2017). Internet of Things: Visions, technologies, and areas of application. Automation, Control and Intelligent Systems, 5(6), 83–91. doi: 10.11648/j.acis.20170506.1.
Akanksha Bali, Mohita Raina. (2018). Study of various applications of Internet of Things (IoT). *International Journal of Computer Engineering & Technology (IJCET)*, 9(2), 39–50.
A. Murray, M. Minevich, A. Abdoullaev. (2011). The Future of the future: Being smart about smart cities. *KM World*, http://www.kmworld.com/Articles/Column/The-Future-of-the-Future/The-Future-of-the-Future-Being-smart-about-smart-cities-77848.aspx.
Andrew Meola. (2020). How smart cities and IoT will change our communities. *Business Insider*, https://www.businessinsider.com/iot-smart-city-technology?IR=T, accessed August 20, 2020.

2 Biomedical Applications Using IoT

Neeraj Kumar Jain, Preeti Mittal, and Rajesh Kumar Saini

CONTENTS

2.1	Introduction		23
	2.1.1	According to Wikipedia	23
	2.1.2	According to Techopedia	23
	2.1.3	According to Trendmicro.com	23
	2.1.4	According to Webopedia	23
2.2	History of IoT		24
2.3	How Does IoT Work?		24
2.4	Advantages of IoT		24
	2.4.1	Communication	25
	2.4.2	Information	25
	2.4.3	Automation with Control	25
	2.4.4	Monitoring	25
	2.4.5	Time	25
	2.4.6	Money	26
	2.4.7	Efficient and Time-Saving	26
	2.4.8	Improved Lifestyle	26
2.5	Disadvantages of IoT		26
	2.5.1	Compatibility	26
	2.5.2	Complexity	27
	2.5.3	Privacy/Security	27
	2.5.4	Safety	27
	2.5.5	Staff Downsizing and Unemployment	28
2.6	How Big Is IoT?		28
	2.6.1	Did You Know?	28
2.7	IoT Applications		29
	2.7.1	Smarter Homes	29
	2.7.2	Wearables	30
	2.7.3	Smart City	30
	2.7.4	Smart Electricity Grids	30
	2.7.5	Industrial Internet	30
	2.7.6	Connected Cars	31
	2.7.7	Connected Healthcare	31
	2.7.8	Smart Retail	31

		2.7.9	Smart Supply Chains	31
		2.7.10	Smart Farming	32
2.8	IoT in the Biomedical or Healthcare Sectors			32
2.9	Why Healthcare?			33
2.10	Applications of IoT in Healthcare			33
		2.10.1	Remote Healthcare/Patient Monitoring	33
		2.10.2	Wearables	33
		2.10.3	Patient-Specific Medicines	33
		2.10.4	Equipment Availability and Maintenance	33
		2.10.5	Medical Asset Monitoring and Better Patient Care	34
		2.10.6	Data Management	34
		2.10.7	Smart Beds	34
		2.10.8	Diabetes Management	34
		2.10.9	Help for the Paralyzed	34
		2.10.10	Senior Monitoring	34
2.11	Biomedical Devices and Apps in IoT			35
		2.11.1	Myo	35
		2.11.2	Zio	35
		2.11.3	MyDario	35
		2.11.4	SleepBot	36
		2.11.5	GOJO	36
		2.11.6	Weka	36
		2.11.7	Apple App to Manage Depression with Smart Watch	36
		2.11.8	Health Application by Intel	37
		2.11.9	YuGo	37
		2.11.10	Sensimed	37
		2.11.11	MoMeKardia	37
		2.11.12	Quell	37
		2.11.13	SmartPump	38
		2.11.14	Withings Thermo	38
		2.11.15	Trak	38
2.12	Benefits of IoT in Healthcare			39
		2.12.1	Economic Benefits	39
		2.12.2	Reduced Risk of Errors	39
		2.12.3	Solution to Problem of Distance	39
		2.12.4	Better Treatment Outcomes	39
		2.12.5	Better Disease Control	39
		2.12.6	More Trust Towards Doctors	40
		2.12.7	Easy Control of Medicines	40
		2.12.8	Maintenance of Connected Devices	40
2.13	Drawbacks of IoT in Healthcare			40
		2.13.1	Privacy Issues for Patients	41
		2.13.2	Accidental Failures	41
		2.13.3	Malware	41
		2.13.4	Lack of Encryption	41
2.14	Conclusion			42
References				42

2.1 INTRODUCTION

The Internet of Things, or IoT, is an interconnection of smart electronic components with others like sensors, networks, and software components, allowing these objects to gather, send, and receive information among themselves using a network [1].

Let's explore different definitions and opinions offered by scholars about IoT.

2.1.1 According to Wikipedia

IoT or Internet of things is the smart advancement of connectivity among household electronic physical devices through interconnection of electronics, Internet, and special hardware like sensors devices can exchange information and talk to each other over the network, which is monitored and controllable remotely. [2]

2.1.2 According to Techopedia

IoT or Internet of things is a new concept of computation that explores the possibility of connecting various physical objects to the internet and so that data and information can be exchanged. IoT is closely making use of techniques like RFID as the medium of communication, but it may make optimal use of sensors, QR codes or other wireless mediums etc. [3]

2.1.3 According to Trendmicro.com

IoT or Internet of things may be described as the next step of the internet and other network inter connections to different sensors and devices—or may be even domestic appliances like lightbulbs, microwave oven, air conditioner, locks, etc. [4]

2.1.4 According to Webopedia

IoT is abbreviation of "Internet of Things", which means continuously expending network of smart objects connected through to establish communication among them using unique identification through IP address over networks. [5]

IoT is a system of arranged computing devices, mechanical and digital machines, objects, animals or anything that has been given unique identifiers (UIDs) and also the ability to exchange data and information over a network without requiring human-to-human (HHI) or human-to-computer interaction (HCI).

IoT can comprise a range of different objects; for example, it can be a patient equipped with cardiac monitoring equipment; an animal with silicon chips fitted; an automobile with sensors attached to alert and inform the driver when the air pressure in the tires is low; a refrigerator with a sensor which informs users that there is no milk left and reorders the specified quantity; or anything else that can be part of a network using IP addresses to transfer information over a network.

Every day, flourishing and growing industries are making optimal use of IoT applications to operate better and more efficiently, to give their customers the best possible service, and to make decision-making as effective as possible, hence making their businesses more and more profitable.

2.2 HISTORY OF IOT

Kevin Ashton, co-founder of Auto-ID Center, MIT, first coined the term IoT or Internet of Things in a presentation for Procter and Gamble about the use of radio frequency identification (RFID) in 1999; he called his presentation "The Internet of Things" because of the then-emerging Internet.

In his 1999 book *When Things Start to Think*, Neil Gershenfeld, a professor at the Massachusetts Institute of Technology (MIT) did not use the term Internet of Things but did give an idea of the emerging IoT. Although Ashton was first to coin the term IoT, the idea of connected devices had originated in the 1970s with the concepts of *embedded internet* and *pervasive computing* [6].

The evolution of IoT is the result of emerging and efficient wireless technologies, microelectromechanical systems (MEMS) and services, and the enormous capabilities of the Internet.

The first Internet appliance was a Coke machine at Carnegie Mellon University in the early 1980s, which was connected to the Internet and alerted programmers about the availability of cold drinks in the machine.

IoT has its origin in machine-to-machine (M2M) communication, in which, without any human intervention, machines can interact with each other over a network. M2M means to connect devices to collect and process data in a controlled environment.

IoT is an extension of M2M with sensors and billions of other smart devices that connect humans, machines, and applications to gather and process exchanged data.

SCADA (supervisory control and data acquisition), used for process control, is also important to discuss in this regard. It uses hardware and software components to collect data from remote and distant locations to control devices in varied conditions. Hardware collects and stores data, while SCADA software installed on it is used to process data. First-generation IoT systems can be considered an evolution of late-generation SCADA systems.

2.3 HOW DOES IOT WORK?

IoT is an interconnection of many entities. Primarily, IoT architecture is made up of:

- Networkable smart components equipped with embedded processors
- Sensor components
- Hardware components for communication

As seen in Figure 2.1 collectively these components gather, exchange, and process data collected from the surroundings. These devices exchange the data to be analyzed locally or through cloud services. Connected devices interact with their peers to react on the basis of the data and information. As these devices can operate without much human intervention while people are able to control them for instructions or to fetch data, they are considered to be "smart" [7].

2.4 ADVANTAGES OF IOT

IoT has proved to be very useful in many areas. Let's discuss its many advantages.

Biomedical Applications Using IoT

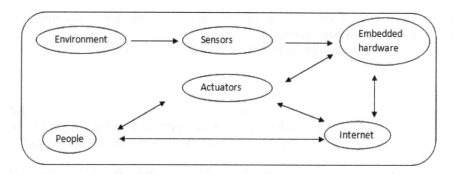

FIGURE 2.1 IoT architecture.

2.4.1 Communication

As IoT is another name for M2M communication, IoT encourages interaction among connected devices. These connected devices are thus capable of maintaining transparency with fewer inefficiencies and high quality.

2.4.2 Information

More data with rigorous processing helps us make better decisions, from simple things like working out what you need to buy at the shopping mall to determining if your company is maintaining a balance between demand and supply. We are surrounded with data but it can produce precious knowledge proportional to direct gain.

2.4.3 Automation with Control

As these physical, connected devices are controlled digitally over wireless networks, there is massive scope for automation under a controlled environment. Machines are now capable of interacting with each other, which results in quicker output.

2.4.4 Monitoring

Monitoring is another of the most obvious advantages of IoT. Prior knowledge of the quantity of something required or the quality of air at a particular place may furnish better information more easily than before. Awareness in advance that you are running out of petrol or that the air pressure in your vehicle's tires is low may make your long drive easier, or knowing that there is a shortage of milk in refrigerator could save you a trip to the store by placing an order for milk to be delivered. In addition, monitoring and controlling expiry dates of items will be easier and no doubt will improve safety.

2.4.5 Time

If we can monitor many things and take appropriate and timely action, this will obviously save a significant amount of time. Frequent visits to the doctor can be avoided, saving time. You can operate your air conditioner and other connected equipment through your smartphone, saving time.

Time is precious and time is money, so IoT can be the best choice.

2.4.6 Money

Time is money and IoT is quite cost-efficient so this is a significant advantage of IoT. The use of IoT will become widely accepted and appreciated if the price of tagging and monitoring equipment is lower than the money and resources needed to establish a traditional system.

By adopting this latest innovation, the vitality and capabilites of such smart gadgets can be efficiently utilized. Care must be taken to avoid bottlenecks, hardware failures, network breakdowns, or physical damage. Thus, we can save money by making efficient use of this innovation.

It automates and controls tasks that need to be accomplished regularly, preferably without human intervention. M2M communication ensures uniformity and quality of services and transparency in their applications. Nevertheless, special arrangements should be made for emergencies.

2.4.7 Efficient and Time-Saving

The M2M interaction results in more efficient applications. More accurate results can be obtained quickly. This means that important and valuable time saved on a regular basis can be utilized to carry out better creative work rather than doing repetitive tasks regularly.

2.4.8 Improved Lifestyle

All such smart applications of this innovation of the IT era result in more comfortable, quicker, easier, and more efficient management of tasks, and hence an improved lifestyle.

2.5 DISADVANTAGES OF IOT

Although this revolutionary technology has many advantages, "nothing is perfect". Some significant disadvantages of IoT are as follows.

2.5.1 Compatibility

At present, due to lack of international standards or protocols for tagging and monitoring devices and equipment, issues may arise about the most efficient way to implement IoT, but this is easily handled. The companies producing the equipment simply need to establish a standard, such as Bluetooth, USB, or RFID.

Different manufacturers will contribute devices of different types to the network. In this situation, issues relating to the compatibility of tagging and monitoring devices will obviously arise. This problem can be solved easily, however, if manufacturers are obliged to follow the same standards and protocols.

2.5.2 Complexity

All very complex systems are at significant risk of shutting down due to bottlenecks or failures. Hence, there is always the possibility of failures in the case of IoT as well. The following two examples show the repercussions that can occur in the case of failure of this complex system.

Case 1: If a message is received by several members of the same family saying that their milk has expired, then there is a chance that everyone might purchase milk. This will not only result in a waste of time and money but will also cause annoyance and frustration..

Case 2: An error in the software component may mean that it automatically keeps ordering a battery for a laptop every hour for a couple of days or weeks or after each restart, while all that was needed was a one-off replacement.

As IoT is a complex network of a heterogeneous nature, and any minor or major failure or errors may mean significant losses. Even a simple power failure for a short time may result in inconvenience or, at worst, destruction.

2.5.3 Privacy/Security

In order to establish communication among devices over the Internet, a huge amount of data is transmitted and received, creating ample opportunities for hackers. Issues related to encryption and decryption must be taken very seriously.

Nobody wants his/her relatives or employers to know about his/her illness, or medications prescribed, or how good or bad his/her financial situation is.

2.5.4 Safety

In IoT, all the connected equipment, such as household electronics, water supply, transport, and industrial machinery, as well as countless other devices, share a lot of information. This information is vulnerable to attack by unauthorized users with malicious intent. It may be very dangerous if private and confidential information like financial details are accessed by hackers.

These dangerous consequences could include:

- A notorious hacker changes your medical treatment regime.
- A department store sends you an alternate product that does not suit you, food of a flavor which you dislike, or a product which is expired or is about to expire.

Hence, safety is ultimately in the hands of the consumer to verify this information thoroughly.

2.5.5 Staff Downsizing and Unemployment

With this automation of daily activities, there will be less need for human resources like workers and less skilled staff.

Such people may lose their jobs as a result of the automation of routine activities. This will result in an unemployment problem. Such a situation can arise with any emerging technology but can be resolved by providing training in newer skill sets required for handling the new challenges that come with IoT [8].

2.6 HOW BIG IS IOT?

This latest era of connectivity is now expanding its reach beyond computers and smartphones. It is now connecting automobiles, making homes smarter, facilitating connected wearables, developing advanced cities, and creating intelligent healthcare equipment. The result is that people's lives will now be totally interconnected. According to a report by Gartner, by the year 2020, the number of connected devices will reach 20.6 billion.

A survey conducted by HP reveals the enormous growth of connected smart devices in IoT over the years, as shown in Table 2.1. It seems that we are heading towards an automated and smarter world.

These connected smart devices are a way for the physical and digital world to improve the standard of living and productivity of individuals, societies, and industries.

A survey carried out by KRC Research in Germany, the United States, the United Kingdom, and Japan has revealed that customers are more interested in adopting elements of the smart household, like smart fridges and thermostats, and so on, to change the way they use these things [9]. People's opinions on which connected device they were most likely to use in the future, as recorded by the survey are shown in the graph in Figure 2.2.

2.6.1 Did You Know?

- ATMs are assumed to be one of the first IoT devices, having gone online back in 1974.
- In 2008, there were more Internet connected devices than the population of the planet.

TABLE 2.1
Increasing Connected Devices

Year	No. of Expected Connected Devices
1990	0.3 Million
1999	90.0 Million
2010	5.0 Billion
2013	9.0 Billion
2025	1.0 Trillion

Biomedical Applications Using IoT

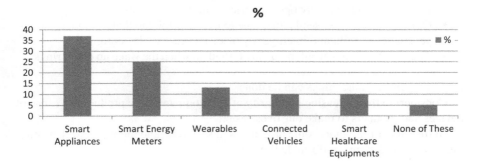

FIGURE 2.2 Connected devices and the percentage of survey participants who chose each device as the one they were most likely to use in the future. Source: GMSA Report.

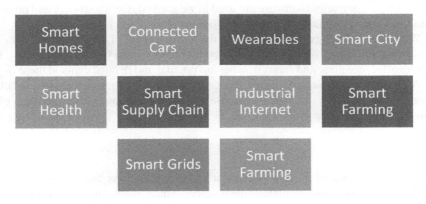

FIGURE 2.3 Applications of IoT.

- In year 2015, there were more than 4.9 billion (4,9000,000,000+) connected devices in the IoT. The number of smartphone users topped 6.1 billion (6,100,000,000+).
- It is predicted that by the year 2020, the number of IoT connected devices will reach 50 billion. 25% of vehicles will also be connected to the Internet [10].

2.7 IOT APPLICATIONS

The main purpose of IoT is to automate human life. In Figure 2.3, some very important applications of IoT are shown which will truly achieve this goal.

2.7.1 SMARTER HOMES

On Google, "smart home" has become the most searched IoT-associated term. Everyone would like to live in a smart house where:

- We could switch on the air conditioner 15 minutes prior to reaching home.
- Lights turn on/off automatically whenever we enter or or leave the house.

- Doors can automatically open for friends and remain closed to strangers.
- Our refrigerator can send a message to our smartphone that its time to buy vegetables or milk and intelligently place an order to the store.
- The hot water can get automatically turned on when the wake up alarm rings.

And many more smart things can be done.

Considering the increasing popularity of IoT, companies are producing a range of devices and products to make our lives easier.

At the same time, smart home products will save time and money as well as offering a better and more comfortable lifestyle.

2.7.2 Wearables

Sensors with compatible software are used in wearable devices to collect and process data from users, ultimately gaining essential insights.

Wearable technology is a milestone in the journey of IoT. This is perhaps is one of the earliest industries which are taking maximum advantage of IoT. Devices like Fit Bits, cardiac monitors, and smart watches are popular and in huge demand these days.

A Guardian glucose monitoring device is also a popular wearable which helps people suffering from diabetes. A glucose sensor injected into the skin detects glucose levels in the body.

2.7.3 Smart City

Smart cities are one of the most sought-after IoT applications. Smart law and order, road transportation, power management, surveillance, water supply, pollution control, ground water management are some examples of smart applications of IoT for smarter cities. Products like smart trash cans, for example, can inform municipal corporations whenever they need to be emptied.

By using smart sensors with web applications, people can find vacant parking spaces in the city. When connected to a smart electricity system, sensors can detect meter tampering, frequent faults, and so on.

2.7.4 Smart Electricity Grids

Future power grids for electricity will be reliable smart grids. Smart grids will collect data automatically, allowing services for electricity users and suppliers to be improved through monitoring and control.

These smart grids will now be able to detect and report electricity theft or misuse in a timely manner and can provide error-free services. Anything from automatic bill generation to fine management can be done easily.

2.7.5 Industrial Internet

The Industrial Internet or Industrial Internet of Things (IIoT) is the new sensation in the industrial sector these days. IIoT is helping this sector with sensors, connected devices, software, and techniques for data analytics to produce smarter machines with smarter methods.

Performance-wise, these smart machines are likely to be more accurate and reliable than human beings. Companies can analyze this collected data quickly to check for inefficiencies and other issues.

IIoT has enormous potential for maintaining quality and supportability. Operations like merchandising; sharing current data about stock among suppliers, providers, and retailers; and conveyance facilities will build a more efficient production network.

2.7.6 Connected Cars

IoT has a vital role to play in intelligent transportation systems.

IoT, being a digital technology, has been emphasized in the automation of vehicles for smart and efficient functioning.

A connected car behaves smartly to enhance the performance of operations like comfort, entertainment, and maintenance using sensors and internet connectivity.

Major big companies like Google, BMW, and Apple will soon be announcing innovations in automobile production.

2.7.7 Connected Healthcare

IoT applications can transform traditional medical systems into proactive wellness-based smart medical systems. Traditional processes of healthcare or nursing will give way to smarter treatment systems.

Existing medical research practices are working on limited resources in a controlled environment. IoT explores innovative ways to use an ocean of important data through analysis, real-time data, and diagnosis.

Smarter devices used in IoT are far superior in terms of power consumption, accuracy, precision, and availability.

2.7.8 Smart Retail

In a retail store, many tasks can be done more efficiently with the help of this ubiquitous technology. Tasks such as foot-traffic monitoring, equipment maintenance, demand alerts, and warehouse maintenance can be carried out and monitored very efficiently on a regular basis. Other functions which can help the retail sector are:

- Enhanced supply chain management
- Better customer service
- Smarter inventory management
- Automated checkout

2.7.9 Smart Supply Chains

Using IoT-connected devices, RFID tags with positioning sensors may help wholesalers and retailers to obtain precise information about the location and movement of ordered items from producers to customers, even down to details such as when order was placed at the store, the time taken to transport the products, and the required suitable temperature at which the product was being stored. Any issues can be addressed and resolved quickly in a systematic way. For example, conside products with a short expiry date that require a fixed temperaure for storage; the data provided by RFID tags will certainly be useful in monitoring such perishable items [11].

2.7.10 SMART FARMING

Smart agriculture or smart farming is an extension of information and communication technologies (ICT) in agriculture, representing a smart and hi-tech way of growing quality food products in a hygienic and sustainable way.

IoT-based smart agriculture is much more efficient than conventional methods of agriculture. Farmers can now monitor their fields and crops from anywhere via their smart devices.

Sensors for light, temperature, moisture, humidity, and so on help in automating farming and irrigation systems. This IoT-based, tech-savvy approach to agriculture is exceptionally efficient and productive in comparison to traditional methods. Moreover, smart farming is also beneficial for the environment, as it allows for optimal consumption of water, fertilizers, and other essential resources.

2.8 IOT IN THE BIOMEDICAL OR HEALTHCARE SECTORS

The most important and ubiquitous impact of IoT will probably be in healthcare and medical science. By 2020, it is predicted that the healthcare sector will account for 40% of the use of IoT technology, which is more than any other sector, forming a $117 billion market [12]. This amalgamation of healthcare, computer information and ICT, such as healthcare informatics or electronic-health, will definitely revolutionize healthcare by reducing medical expenses and inefficiencies, and saving lives.

This revolutionary technology of connected devices (IoT) can transform a typical old-fashioned hospital into a smart IoT hospital. All a heart patient has to do is present a smart card, and hospital staff can fetch all his or her details from a safe cloud by scanning the card. With a few clicks, hospital staff and doctors can access the patient's old prescriptions, lab diagnosis results, and so on [13, 14].

With the advent of the latest healthcare techniques, with robotic consultants and surgeons available in good hospitals, people can access a whole new generation of high-quality healthcare within the quickest possible timeframe. These emerging technologies are not only revolutionizing patient care but also helping doctors to work more efficiently. A survey has shown that the US Department of Health and Human Services will save approximately US$ 300 billion due to medical advancements in which IoT plays a vital role.

IoT can facilitate an integrated set of connected medical equipment to help both medical practitioners and patients, improving treatments and making them easier and more comfortable. New developments along with specialized hardware such as sensors help make an IoT-enhanced healthcare field a reality.

Biomedical Applications Using IoT

2.9 WHY HEALTHCARE?

Healthcare is one of the most challenging areas in which IoT plays an important role. It has enormous potential to provide very sophisticated healthcare services for the benefit of humankind. It covers all medical aspects from individual patient treatment and population health monitoring to self-care and social servicing. It is exploring astonishing methods of disease prevention, cure, and prediction as a result of close monitoring, controlled treatments, and timely diagnosis of ailments.

These smart IoT devices and equipment are capable of storing and processing medical data on a continuous basis from patient to patient; they can detect patterns, make predictions, and prevent progression of diseases in a timely manner.

2.10 APPLICATIONS OF IOT IN HEALTHCARE

IoT is capable of improving existing medical practices to a great extent. It can make great contributions in establishing smart systems for better patient care and more systematic patient-oriented treatment. Some innovative IoT applications in healthcare industry are as follows.

2.10.1 REMOTE HEALTHCARE/PATIENT MONITORING

Healthcare equipment and devices connected to IoT and specialized software allows doctors to access patient data from their medical records online or offline. Regardless of the physical presence of the patient, doctors can now carry out analysis, give advice, and send notifications to patients and other stakeholders.

2.10.2 WEARABLES

These smart devices are designed to continuously monitor patient activity. They can be worn on the wrist, chest, or other parts of the body. They are smart enough to record the number of steps a person has taken, the calories they have burnt, their blood pressure, their heart rate, and much more. In emergencies, these smart devices can be life-saving.

2.10.3 PATIENT-SPECIFIC MEDICINES

Specialized smart devices and IoT healthcare sensors are able to offer medical assistance; a qualified doctor can use them to deal with the problems and requirements of individual patients.

2.10.4 EQUIPMENT AVAILABILITY AND MAINTENANCE

IoT can ensure the availability of major healthcare devices and make sure all medical equipment is in good working order as patients may need it at any time. It keeps track of all connected devices, which helps to prioritize fixing any functionality issues.

2.10.5 MEDICAL ASSET MONITORING AND BETTER PATIENT CARE

As this technology allows most healthcare tasks to be performed by devices, doctors and other medical staff will have more and more time to take care of their patients. Instead of spending more time on searching and fixing things in the traditional system, now they can focus more on their patients. IoT and healthcare are thus mutually beneficial.

2.10.6 DATA MANAGEMENT

IoT offers a number of ways to collect patient information for hospitals both on-site and outside of the hospital. Healthcare uses telemetry to collect and communicate data automatically and remotely. This allows doctors and other medical practitioners to take care of their patients promptly and provide them with better treatment [15].

2.10.7 SMART BEDS

For patients who cannot adjust their bed positions themselves, smart beds are a very useful solution. This IoT tool can sense when the patient tries to move but is unsuccessful. It works smartly by correcting the bed orientation, height, or pressure to make the person more comfortable. This eliminates the need for a nurse to be available all the time; it is thus not only money saving but also provides immediate service for patients.

2.10.8 DIABETES MANAGEMENT

Diabetes is one of the most common diseases worldwide. Therefore, IoT developers have created mobile apps to support patients and help them monitor and control it on their own. These apps include all kinds of sensors and equipment which can track glucose levels in the body and alert patients to take insulin. Some of the latest wearables are themselves capable of injecting an insulin dose, according to requirements.

2.10.9 HELP FOR THE PARALYZED

Patients suffering with paralysis need special attention and more care than other patients as they require continuous assistance. With IoT, specialized devices are used to detect the needs of such patients through facial-expression recognition or analysis of their thoughts through headset devices. This helps such patients express their wishes. While this technology helps nurses to take care of patients who need constant attention, it will never replace the need for nurses in hospitals.

2.10.10 SENIOR MONITORING

IoT is serving the healthcare industry not only through direct medical action but also through simple monitoring and sensing. It is always difficult to track elderly patients, who are often weak and face multiple problems in carrying out daily routine tasks. IoT gadgets can assist relatives and/or hospital staff to take care of seniors.

2.11 BIOMEDICAL DEVICES AND APPS IN IOT

In this era of information, data and knowledge are the keys. This is the time when customers are eager and curious about what they actually want. *myTomorrows* is one very useful website depicting a rapidly revolutionizing business scenario in which people have direct access to medical services [10].

Blood-glucose monitors, cardiac monitors, and other devices incorporating smart sensors are predominantly in use, and more and more companies are developing apps and software to improve healthcare and to regulate digital health tools. Some of the most popular IoT biomedical devices used for healthcare in this new age are listed in this section.

2.11.1 Myo

This is a movement controller normally used in games, but it is now being used in orthopedics to monitor and guide patients doing exercise after bone injuries. It helps patients to monitor their progress and also helps doctors, as they can measure the type and angle of movement and advise their patients accordingly.

The Myo Armband is a gesture-control armband which uses Bluetooth technology that captures muscle activity and gives the user touch-free control using with hand movements, gestures, and motion.

2.11.2 Zio

The Zio Patch is used to measure heart rate and electrocardiograms (ECG). The Zio patch can give truly uninterrupted indications without the need for cables or maintenance, and gives analyzable data. Zio is recommended by expert cardiac technicians to give give better, clearer, and more comprehensive insights into a patient's heart activity.

2.11.3 MyDario

MyDario is a very useful blood-glucose monitoring system with the following features:

- It is a smart pocket-sized or hand-held device.
- Only the smallest drop of blood is required.
- It can be used by anyone.
- It can be used anywhere, at any time.
- It gives quick and accurate measurements.
- It works rapidly, giving results in less than 6 seconds.
- It involves no wires, batteries, or hassle.
- It automatically records blood-glucose levels.
- It analyses the patient's condition.
- It keeps track of the patient's diabetes metrics, medication, and physical activity.

- Current and accurate information and personalized reports can be shared with family members and doctors.
- It has an in-built hypo alert with GPS location for emergencies [16].

2.11.4 SleepBot

SleepBot is a multi-functional sleep app, but it is basically used to track sleep patterns. It works as a "smart alarm" and awakens people more gently using this feature. SleepBot can be used by those who need to keep track of their sleep and improve their sleep, and who want to wake up fresh and in a more comfortable way [17].

Many IoT healthcare projects in the form of mobile apps and websites are already available for use these days. Let's look at some successful and popular examples.

2.11.5 GOJO

In the medical field, especially in intensive care units (ICUs), there is always a high risk of infection. Washing your hands and being free from germs and bacteria is very important. So proper hand washing is the most efficient way to reduce intra-hospital infections. GOJO, a company which usually supplies antibacterial gel and other antibacterial solutions for hands, is now using the Azure IoT Suite from Microsoft, a system comprising sensors and monitors for maintaining hygienic procedures in hospitals. It can observe, monitor, and prepare reports on levels of compliance with rules and regulations. The main objective of this project is to enhance the level of personal cleanliness of patients, doctors, and other attendants in hospitals with IoT healthcare solutions. This has proven to be safe and to save lives.

2.11.6 Weka

For the transportation and storage of different vaccines, physicians have to make special arrangements. To overcome this issue, Weka Solutions is the company which took the initiative and is working with GOJO and Azure IoT to develop smart refrigerators. With this technology, a normal refrigerator can be utilized as a smart device for improving the monitoring and maintenance of important vaccines storage for timely and safe transportation. This continues to work even in areas of countries and cities where there is no proper power supply available. This IoT healthcare technique is considered to be an important and very useful invention.

2.11.7 Apple App to Manage Depression with Smart Watch

Experts from Cambridge Cognition have collaborated with pharmaceutical giant Takeda, to develop the Apple Watch app, which helps in analyzing behavior and cognitive symptoms in patients battling depression.

According to a recent survey, depression is major cause of disability and affects 350 million people worldwide. By combining specialized wearables with proven solutions in neurosurgical science, experts and scientists have developed apps that

Biomedical Applications Using IoT

gather important active and passive data about the psychological conditions of patients. Regularly collected data regarding the routine activities of patients can help specialists improve their understanding, make accurate decisions, and achieve results in a timely manner. This is certain to improve medical treatment processes.

The Apple Watch app, used to manage depression, is quite popular today.

2.11.8 HEALTH APPLICATION BY INTEL

Intel and the Flex supply chain corporation have created IHAP (Intel Health Application Platform), a new IoT system. Its purpose is to take care of patients remotely even without using a tablet or smartphone. This platform uses a smart device that helps doctors to get in touch with patients remotely with the help of proprietary software from Intel. All the information is stored in the cloud to be accessed by doctors as and when required.

2.11.9 YUGO

This Israel-based startup and Microsoft biogaming product is basically a physical therapy system that uses Kinect. Used by physiotherapists to help their patients, it not only keeps track of a patient's movements but also automatically maintains records of their therapy.

2.11.10 SENSIMED

Switzerland-based firm Sensimed has developed smart contact lenses named Triggerfish that assist doctors in monitoring the spread of glaucoma. A pair of these contact lenses equipped with sensors can continuously store ocular dimensional changes throughout the day. This record from the lenses is saved wirelessly to a recorder in the patient's neck. After a certain time interval, this data is automatically sent from the recorder to the doctor via Bluetooth.

2.11.11 MOMEKARDIA

Lowell, Massachusetts-based company InfoBionic has developed and delivered a patient monitoring system called MoMeKardia, which is designed to help heart patients by detecting cardiac arrhythmias by collecting data on ECG, respiration, and motion. This lightweight monitoring device is designed to be worn as a collar or a belt.

2.11.12 QUELL

Neurometrix has designed a new smartphone-based device called Quell, a painkilling wearable. This device helps to treat chronic pain by sending signals to the brain using nerve stimulation. This useful device, which can be worn on wrist, allows users to directly control the level of nerve stimulation.

2.11.13 SmartPump

A California-based startup has developed a smartphone-connected breast pump, which uses a hydraulic system instead of air to pump breastmilk. Pumping is improved as well as being comfortable and noiseless.

2.11.14 Withings Thermo

Withings, along with Nokia, have developed Withings Thermo, a Wi-Fi-enabled thermometer. Instead of being put under the arm or tongue of the patient, it senses and records their temperature from their forehead.

2.11.15 Trak

Sandstone Diagnostics has developed Trak, an app-connected simple home test to check male fertility. This device offers a more sophisticated, less awkward, and less inconvenient testing method. The device uses a centrifuge to separate sperm cells into specifically designed cartridges. It analyzes the samples and transmits a diagnosis to a mobile app, informing the user whether his sperm count is low, moderate, or optimal according to World Health Organization (WHO) guidelines.

Where is the healthcare industry and pharma in all this chaos? It is interesting to notice that symptoms and need of advancements in healthcare is crossing barriers of its old medicine-centric culture and processes.

Let us consider some interesting scenarios.

- J and J and Google have come together to perform robotic surgery using artificial intelligence and machine learning. For wearable devices like blood-pressure monitors and so on, they are working with Philips [12].
- Novartis and Google are working together on sensor technologies to produce smart lenses and other wearable devices to monitor blood-glucose levels [13].
- Glaxo is to invest in electroceuticals, "bioelectrical" drugs that operate by micro-stimulation of nerves [14].

These sensors are useful in collecting data from their surroundings, and this data can be transformed into meaningful information to guide medical practitioners and to support pharmaceutical development. Many useful sensors are now being used in the biomedical and healthcare sectors as well. Body sensors originally designed to be used primarily by athletes are now increasingly popular in heterogeneous markets, but their largest contribution is in the biomedical sector. They will allow consumers and the pharmaceutical industry to share not only important diagnoses regarding such things as blood sugar, pulse, blood pressure, ECG, and so on, but also more sophisticated information like facial expressions, sleep patterns, and heartbeat patterns.

It has been forecasted that in the future, people will use their smartphones and watches to monitor health issues more frequently than they do now to check their emails or social platforms. Consider an elderly person who is recovering from an

Biomedical Applications Using IoT

illness at home; a range of medical devices can collect data from his surroundings, which can be rapidly accessed in an emergency [15].

2.12 BENEFITS OF IOT IN HEALTHCARE

The use of IoT in the healthcare sector is vital and is becoming more and more popular every year. Some of the visible and widely accepted benefits of IoT are as follows.

2.12.1 Economic Benefits

Very economical sensors and other cheaper special hardware devices have brought the cost of treatments down to a point where they are significantly lower than traditional treatments. This technology helps medical practitioners and specialists to take care of patients in real time. The physical presence of patients is no longer required, avoiding the costs involved in a personal visit. Admission to nursing homes and hospital stays can now be avoided, also saving money. It is also economical for doctors, who can avoid costly visits by monitoring the health of their patients online. Moreover, instead of appointing nurses and other medical staff for the monitoring of patients in ICU, they can now be replaced by connected cameras and alarms.

2.12.2 Reduced Risk of Errors

Computerized data collection and processing leads to data-driven decisions, which tend to produce more accurate results and more satisfactory treatment. A reduced risk of mistakes in diagnosis and treatment make procedures more reliable for patients.

2.12.3 Solution to Problem of Distance

Some countries have already adopted telemedicine systems, which can be used to effectively provide better treatments by facilitating a conversation between patient and doctor. A quick communication between doctor and a patient using telemedicine can help a lot to the doctor for general examination of the patient.

In Singapore, telehealth technology has revolutionized the healthcare industry. Doctors can now remotely monitor patients living in distant and rural places, no longer requiring them to attend a hospital.

2.12.4 Better Treatment Outcomes

With the IoT-based health monitoring systems comprising connected medical devices and cloud computing, medical practitioners are always able to monitor their patients' current information in real time. This helps a lot in making accurate and timely diagnoses, leading to better treatment outcomes.

2.12.5 Better Disease Control

With IoT and more sophisticated technologies, patients can always be treated under close supervision. For doctors, it is important that patient data is available when it is needed. By receiving updated information regularly, doctors can detect diseases in the early stages and can start treatments in plenty of time.

Due to the advent of IoT and big data in medical and healthcare, data can now be gathered from remote areas, while previously it was quite difficult and had to be done in person by doctors. Smart breath analyzers and thermometers can now collect current data and store it to global healthcare and medical systems; doctors and technicians can then quickly analyze a patient's condition and share it using disease monitoring and controlling tools in place at remote locations.

At the Virginia Tech Network Dynamics and Simulation Laboratory, tools like HealthMap and EpiCaster for disease monitoring are combining IoT data with:

- Population data
- GIS information
- Land-use data
- Unstructured data from social media

They also use data from other places and sources to monitor other health risks and threats like Zika, H1N1, and others.

2.12.6 More Trust Towards Doctors

This very useful technology (IoT) in hospitals improves the entire healthcare mechanism. Patients are able to contact their doctors more reliably, and at the same time doctors can also get in touch with patients at any time to offer appropriate advice and treatment. This definitely enhances patients' trust in their personal doctors.

2.12.7 Easy Control of Medicines

IoT in healthcare and all applications and equipment will help doctors to control the supply of medicines, which can be implemented easily with less cost and effort. This way, where a small software component is responsible for monitoring and ensuring supply and demand of medicines, the control process becomes much more accurate and reliable.

2.12.8 Maintenance of Connected Devices

Medical devices are expensive, and equipment requires proper maintenance to perform properly. In the case of the breakdown of any device, IoT can help to detect and repair the fault. This technology will definitely save lives since it will detect and fix all possible problems with the specific devices in a timely manner (see Figure 2.4).

2.13 DRAWBACKS OF IOT IN HEALTHCARE

There is no doubt that IoT is revolutionizing the healthcare and biomedical sectors but on the other hand there are possible significant issues which cannot be ignored. Some serious pitfalls that we should be ready to face are listed in this section.

Biomedical Applications Using IoT

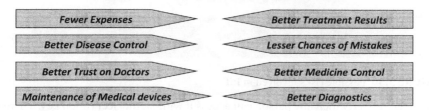

FIGURE 2.4 Advantages of IoT in healthcare.

2.13.1 PRIVACY ISSUES FOR PATIENTS

Privacy is a serious concern. Doctors are dealing with patients' personal information, and their complete history is available as record. Medical history is confidential information. Unfortunately, there is the possibility that hackers can hack this data or misuse a healthcare IoT ecosystem, which, in a worst case scenario, may be life threatening for patients in worst case. Malicious hackers could attack a pacemaker or gain control of other life-saving devices with the intent of causing harm. There may also be chance of future risks for medical records, devices, or wearables. Systems and connected devices in hospitals may comprise older equipment using outdated operating systems and software. In this situation, data security becomes a serious challenge.

2.13.2 ACCIDENTAL FAILURES

In an IoT system for patient health monitoring, a small error may be dangerous or life threatening. No method or even the latest technology can assure a 100% error-free system. As IoT in healthcare works in a connected environment, one tiny failure may result in loss of many lives. This is why it is essential to pay meticulous attention to the functionality of all software development, hardware manufacturing, integration, and combined performance.

2.13.3 MALWARE

As the Internet, which is backbone of IoT in healthcare, is unfortunately full of viruses, Trojans, worms, and so on, that can infect operating systems, healthcare devices, and IoT infrastructure may also be prone to these threats. Consequently, special and effective antiviruses and firewall systems must be used to keep software components updated and protected from undesirable and dangerous attacks.

2.13.4 LACK OF ENCRYPTION

To keep patient records safe from hacking, developers have to emphasize and pay most of their attention customizing encryption methods to protect the system from unauthorized attacks. Sometimes, unfortunately, if systems are not encrypted or

FIGURE 2.5 Possible risks involved in IoT.

protected properly then any person can access it without the decryption key. There is a risk of malicious and unauthorized access to all unprotected data (see Figure 2.5).

2.14 CONCLUSION

As we have seen, IoT is a comprehensive system comprising a number of smart devices that cannot be developed and elaborated overnight. It takes many expansive resources and lot of time for experts to develop really high-grade products. In order to include smart IoT devices in existing systems, one must have access to each and every minute detail about a system to upgrade the product. IoT in the biomedical field requires even more concerted efforts by developers and testers, as a small error may be life threatening for many.

IoT is gaining popularity in the healthcare sector as people are finding it to be a reliable and more accurate technology for monitoring and controlling their health conditions. People can utilize smart IoT devices to keep track of their regular checkups, appointments, variations in blood pressure, calories burnt, heart function, and so on. One of the most appreciated services provided by IoT is remote health monitoring, through which patients are monitored and treated by doctors remotely. Doctors can easily locate patients by tracking the locations of their devices, which is quite time saving and quick. Due to the fact that the smartphone is now within the reach of ordinary people, it is gaining popularity, as people are used to using mobile apps for various tasks. In the healthcare industry, mobile apps can improve interaction between patients and doctors over a reliable and secured medium.

IoT proves that technology is developing at an increasingly rapid pace. Doctors and other medical practitioners are upgrading their services for the betterment of patients. They are looking forward to improving treatment processes using IoT in the healthcare and biomedical sectors. Ultimately, improvements are needed to provide better IT infrastructure and, most importantly, to make people aware about this technology and its effectiveness.

IoT is ready to play a revolutionary role in the approach taken by hospitals. Physicians, surgeons, and IT engineers should work together to achieve the best results.

REFERENCES

1. Zanella, A., Bui, N., Castellani, A., Vangelista, L., and Zorzi, M. (2014). Internet of Things for smart cities. *IEEE Internet of Things Journal*, 1(1), 22–32.
2. https://en.wikipedia.org/wiki/Internet_of_things.

Biomedical Applications Using IoT

3. https://www.techopedia.com/definition/28247/internet-of-things-iot.
4. https://www.trendmicro.com/us/iot-security/Solutions/IoT-Security.
5. https://internetofthingsagenda.techtarget.com/definition/Internet-of-Things-IoT.
6. https://www.slideshare.net/MohanKumarG/internetofthings-iot-aseminar-ppt-by-mohankumarg.
7. https://www.google.com/url?sa=iandrct=jandq=andesrc=sandsource=imagesandcd=andved=2ahUKEwj.
8. https://www.linkedin.com/pulse/advantages-disadvantages-internet-things-iot-tommy-quek.
9. https://www.analyticsvidhya.com/blog/2016/08/10-youtube-videos-explaining-the-real-world-applications-of-internet-of-things-iot/.
10. https://www.researchgate.net/figure/IoT-Roadmap_fig1_331257501.
11. https://www.covisint.com/blog/the-evolution-of-iot/.
12. https://d2h0cx97tjks2p.cloudfront.net/blogs/wp-content/uploads/sites/2/2018/05/IOT-.Applications-and-Usecases-01.jpg.
13. https://www.iotforall.com/retail-iot-applications-challenges-solutions/.
14. https://www.ncbi.nlm.nih.gov/pmc/articles/PMC4981575/.
15. https://www.researchgate.net/publication/306022306_Medical_Internet_of_Things_and_Big_Data_in_Healthcare.
16. https://www.ncbi.nlm.nih.gov/pmc/articles/PMC4981575/.
17. https://datafloq.com/read/8-ways-iot-can-improve-healthcare/3066.

3 Emerging Technological Advances in Healthcare

Sweta Silpa Mohapatra, Ashirwad Kumar Singh, and S. Shiva Koteshwar

CONTENTS

3.1 Introduction 45
3.2 Vision and Engineering of IoT..46
3.3 Current Patterns and Uses of IoT in Medical Services.................................48
    3.3.1 Persistent Monitoring and Healthcare Delivery: The First Step of IoT Healthcare Ladder 50
    3.3.2 What Therapeutic Gadgets Can Be Associated Utilizing the Social Insurance IoT Arrangement? 50
3.4 Case Study for IoT in Healthcare..51
3.5 Difficulties Faced by IoT in Human Services or Healthcare........................56
    3.5.1 How Practical Is It to Actualize IoT Medical Services Arrangements? 58
3.6 The Future of IoT in Human Healthcare 58
3.7 Conclusion 59
References..60

3.1 INTRODUCTION

A system whereby physical devices featuring silicon sensors are interconnected to the Internet, much like a wireless technology, is defined as the Internet of Things (IoT) [7]. This technology has many applications for everyday use; for businesses, such a technology could result in a massive savings if leaky pipes could automatically trigger a sensor without wasting any time [7]. Similarly, another example of the use of IoT is if warehouses could automatically put in an order for stocks that are running low [7], or if an air-conditioner in a house could be operated remotely by the owners from a huge distance away [7]. The speed at which IoT is becoming more popular can be best understood from the fact that the market demand for such devices is expected to double over the span of four years starting from 2017 [8]. The service providers for IoT would mostly be cloud-based [7]. Among the certain roadblocks towards the implementation of IoT are the returns on the technology that is adopted, data security, and synchronization with current technologies [8]. Forbes have made some predictions about how IoT will unfold over the years to come. They conclude that about 85% of firms would be either using or considering using IoT, with the manufacturing sector making a big leap. Many more devices that work on the concept of natural language processing, like that of Alexa and Siri, and many more orally communicating devices will be developed, and we will also see the emergence of a 5G telecom network that will further speed up the process involved in the working of IoT [9]. It is being predicted that before long, around 7 billion IoT appliances could be in existence [7]. There will be a tremendous growth in the production of sensor chips [10] and radio frequency identification tags or RFID. Smart cities will all be using IoT in one way or another [10]. Companies like Fitbit (healthcare), Amazon's Alexa (E-commerce), Tesla (innovation), Uber and Spotify (collaboration) are adapting to IoT technology in the present era.

In this dynamically changing world, modifications, enhancements, and so on are happening so quickly that one cannot afford to miss out on catching up with technology. IoT is one of the emerging technologies that can contribute in a significant way. The best part about IoT is that it enables humans and devices to remain connected anytime, anywhere, with anything and anyone, in one way or another. The real reason why IoT has the potential to be a success is the advanced technologies now available. With the expanding number of gadgets interacting with each other, a huge amount of data is being transferred. The outflow of information should be stored properly and analyzed using data analytic software to get basic data for the clinician and patient [1]. However, only the devices within a given the medical infrastructure are connected, and this framework allows access to clinicians for medical applications.

IoT enables doctors to monitor their patients easily, and IoT-based wearables help doctors to examine patients regularly and can help them identify disease at an early stage. In helps in personal health tracking, chronic disease tracking ,and real-time location tracking. IoT considered to represent the next industry revolution, and it is estimated that the industry will grow by more than 250% in 2019 [12]. Recent advances in the technology helps doctors to track unusual changes like rapid heartbeat, rises in blood pressure, or increase in blood sugar levels; it can also be used to identify the severity of the issue and can thus allow doctors to act in time to save lives. The term 'Internet of Things' was coined by K. Ashton way back in 1999. IoT has many and wide-ranging applications in various domains such as the automobile industry, telecommunications, remote sensing, inventory management, smart home appliances, and so on. One of the fields in which IoT is the most useful is the healthcare industry. IoT enables high quality medical services to be provided at low cost. In the healthcare industry, IoT is used widely in elderly health care through the use of electronic wearables; these are sensing devices which can be worn to monitor heart rate, body temperature, oxygen levels, blood-sugar levels, and so on without affecting normal everyday activities. We can see rapid growth in this sector and many reputable companies are investing in this field [11].

3.2 VISION AND ENGINEERING OF IOT

IoT is still at its nascent stage so everyone speculates about its future differently. There are three visions of IoT which are dependent on the things, the Internet, and semantic points of view [3]. The three views of IoT intersect with one another (as shown in Figure 3.1) to generate the ideal IoT plan.

- Things-directed vision
- Internet-directed vision
- Semantic-directed vision

1. *Things-directed vision*: This first aspect provides the viewpoint that every device can have sensors like those present in fast-moving vehicles, such as RFID tags, NFC, remote advancements, etc.
2. *Internet-directed vision*: This second aspect provides the viewpoint that each one of the physical devices should be connected via the web, which would then be developed by using IP for every related article. This second aspect further builds up the base for data coordination of smart devices, which has to be checked continuously.
3. *Semantic-directed vision*: This third aspect provides the viewpoint that every single piece of the information gathered through the receptors of sensors should be broken down to extract important data. It might be practiced with semantic methods that isolate irrelevant information from significant information with their obtained translation. This third aspect of vision gives the foundation for semantic combination using semantic middleware.

Applications which use IoT can be developed by taking advantage of different technologies, for example, remote correspondences, sensor systems, information handling, and so on. The amalgamation of such innovations in an IoT framework can be seen in Figure 3.2. Here, this architecture is presented as five levels: Sensing, Sending, Processing, Storing, and Mining and Learning.

The functions of each layer are given below [2].

1. *Sensing layer*: Sensing layer consists of sensors for wellbeing parameters of patients. Body temperature, circulatory strain, respiratory rate etc. are the

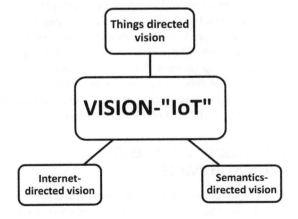

FIGURE 3.1 Three main visions of IoT.

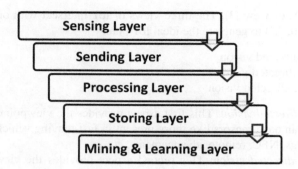

FIGURE 3.2 Structure of IoT in terms of four piers.

most basic parameters utilized. While blood glucose should be estimated for diabetic patients, accelerometers and gyroscopes are being used for managing data in numerous healthcare checking frameworks for anticipating the risk of falls.

2. *Sending layer*: The sending layer helps us to associate and pass on information. Additionally, it can access information from existing IT foundations [2]. In frameworks used for reviewing, wireless technology is utilized for information transmission. Further, these technological measures are useful to guarantee institutionalization and similarity in IoT health systems.
3. *Processing layer*: The processing layer helps, firstly, to gather data from sensing layer; secondly, to share data with storing layer; and lastly to process data after pre-processing it. It consists of preparing parts furthermore, programming applications that apply computational features [2]. The gathered information is prepared for further investigation, creating notes and cautions.
4. *Storing layer*: In IoT-based medicinal services frameworks, the information gathered from the detecting layer is put aside for further analysis [2]. With the development of distributed computing advancements, the responsibility of overseeing and keeping up with the enormous amount of complex restorative information is moved to cloud storage. This creates a marked improvement in the effectiveness and adequacy of the storage and the executives.
5. *Mining and learning layer*: The mining and learning layer help in data mining and machine learning process. These devices are utilized by handling units in storing layer or processing layer, individually, for changing over data to inferences, supporting decisions [2]. Future advancement can use IoT for on-going clinical input as opposed to checking.

3.3 CURRENT PATTERNS AND USES OF IOT IN MEDICAL SERVICES

Prior to IoT, contact between patients and medical specialists was limited to visits and telecommunications. There was no chance that specialists or medical clinics could screen patients' wellbeing consistently and make recommendations as needs be.

Technological Advances in Healthcare

IoT-empowered gadgets have made remote observation in the human services part conceivable, creating the possibility of keeping patients safe and sound, and allowing doctors to provide a superlative service. It has additionally expanded patient commitment and satisfaction, as communications with specialists have become simpler and increasingly effective. Moreover, remote observing of patient wellbeing helps to reduce the length of stays in medical clinics and forestalls readmissions. IoT additionally has a major impact in terms of reducing the cost of health insurance and improving treatment results.

IoT is undoubtedly changing the health insurance industry by revolutionizing the way that gadgets and individuals have interacted. IoT has applications in medicinal services that benefit patients, families, doctors, emergency clinics and insurance agencies.

The standard patterns of the healthcare sector can be interpreted from various perspectives with regards to the innovation, helpfulness, and preferences. Recent advancements in mobile phones and technology have been made with sensors. This brings in the concept of mHealth, which helps us monitor the usage of remote devices in the healthcare industry. IoT applications can also be seen in [6].

1. Tracing of individuals and articles
2. ID and validation
3. Programmed information gathering and detecting

The applications are recorded alongside the use of IoT idea and their advantages.

- *Access to the mobile framework*: This usage relies on the flexible headways that makes the virtual access easy for the current clinical structures. All the framework can be robotized to utilize portable application interface [6]. Examples: Websites, portable applications, etc.
- *Virtual interview*: This application depends on the quality of the network and the sound and video arrangements that help in virtual interviews, training, and treatment strategies [6]. Through this virtual interview process, most of the planned work can happen within minutes or even seconds. There is an opportunity for the development of telesurgery for systems utilizing robots and medical caretaker colleagues.
- *Wireless patient observing*: This application helps monitor basic capacities using patient gadgets [6]. Further, data is being transmitted and shared progressively among people and parental figures in the observation. This is particularly so in the case of chronic conditions like hypertension, diabetes, and so on.
- *Aging set up*: This empowers clinical monitoring for elderly people living independently. [6]. These gadgets help by checking on patients without manual intervention. The information can further be used to send help to the patients. The typical examples of this are video consultations, personal emergency responses systems (PERS), and so on.
- *Medical devices*: This application is used to track illness information and is used as health solution for patient practices, with smart devices being

used to collect data from sensors for further examination by a specialist. Examples include computerized glucometers, blood-pressure devices, pedometers, wearables, and so on.

IoT helps healthcare associations gather information from various sources on a continuous basis and thus to make better decisions. This scenario is changing the human services industry, thereby enhancing its efficacy, cutting costs and leading the way to improved patient consideration.

3.3.1 Persistent Monitoring and Healthcare Delivery: The First Step of IoT Healthcare Ladder

IoT can be used to implement the perfect advanced medical services arrangement that can be of remarkable help to social protection IoT has three possible use scenarios, the primary phase of which utilizes passive observation.

Passive observation arrangements fueled by IoT can use a system of interconnected gadgets, wearables, and smart equipment to gather information continuously and send to an AI. With the assistance of this keen patient observation stage, the burden on therapeutic experts can be decreased, which will eventually improve the nature of social insurance overall. IoT has versatile social insurance applications; wearables and smart devises connected to a system 24×7 can guarantee:

1. Minimization of serious issues due to poor checking or disconnected inheritance framework
2. Improvement of information and data sharing inferable from determined courses, and ongoing sharing by means of IoT arrangements
3. FDA-approved applications controlled by trend-setting innovations to scale up on request

3.3.2 What Therapeutic Gadgets Can Be Associated Utilizing the Social Insurance IoT Arrangement?

IoT restorative gadget reconciliation incorporates a scope of connected smart devices, which can be utilized to monitor patient health and raise issues before the onset of any malady. In the medical services industry, the Internet of Things can be applied to the following:

1. Wearables to constantly screen the wellbeing of patients, take readings, watch designs, and send alarms. This information can be utilized for diagnosis, improved treatment, and upgraded client support.
2. Dynamic passive labels to be worn by the patient or parental figure, and hardware incorporated into the ID cards of emergency clinic staff and patients in order to diminish confirmation time and encourage better administration of assets.

Technological Advances in Healthcare

3. Therapeutic gadgets to follow the utilization of the initial devices for every patient in order to contrast the initial information with gained knowledge and check their status.

3.4 CASE STUDY FOR IOT IN HEALTHCARE

- *Descriptive statistics*:
    Our observation can basically be classified into three sectors:
    1. Doctors (dentists, oncologists, BDS, MBBs students, physicians, MDs, chemists etc.)
    2. Technicians (developers, engineers, technicians)
    3. Sellers (marketing team, sales team, distributors, managers) (most sellers are related to a healthcare equipment manufacturing company)
- *Count of the three sectors*:
    Doctors = 54, Technicians = 8, Sellers = 41 (Figure 3.3 and Figure 3.4).
- *Inferential statistics*:

Hypothesis 1:

H0: There is no significant association between knowledge of IoT and the type of organization in which the employee works.
H1: There is a significant association between knowledge of IoT and the type of organization.

The Chi-Square test was run on SPSS and the output obtained is discussed below.

The first table is basically a cross tabulation obtained for the two variables under consideration and consists of both "Expected" and "Actual" frequency counts for the

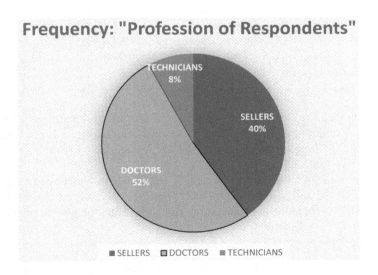

FIGURE 3.3 Pie chart depicting share of professionals.

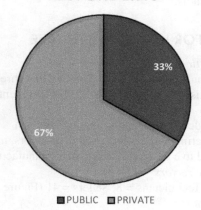

FIGURE 3.4 Pie chart depicting share of respondents on the basis of type of organization.

TABLE 3.1
Crosstabulation Showing "Actual" and "Expected" Frequency between Two Variables "Type of Organization" and "Knowledge about IoT"

Type of Organization and Knowledge about IoT Crosstabulation

			\multicolumn{5}{c}{Knowledge about IoT}	
			1	2	3	4	5	Total
Type of organization	Private	Count	4	9	4	40	6	63
		Expected count	12.2	8.6	4.3	32.4	5.5	63.0
	Public	Count	16	5	3	13	3	40
		Expected count	7.8	5.4	2.7	20.6	3.5	40.0
Total		Count	20	14	7	53	9	103
		Expected count	20.0	14.0	7.0	53.0	9.0	103.0

cross tabulation. The same will be used to carry out the Chi-Square test. The following table is the output of the Chi-Square test specifying the significance level. Also, the notation of "Type of Organization" has two categories: "Private" and "Public" (Table 3.1 and Table 3.2).

Interpretation: As the likelihood ratio is 0.001, which happens to be less than 0.05, we reject the null hypothesis and accept the alternative hypothesis—(H1: There is a significant association between knowledge of IOT and the type of organization).

Hypothesis 2:

H0: There is no significant relationship between knowledge of IoT and experience.
H1: There is a significant relationship between knowledge of IoT and experience.

TABLE 3.2
Chi-Square Test Output

Chi-Square Tests

	Value	df	Asymptotic Significance (2-sided)
Pearson Chi-Square	19.055[a]	4	.001
Likelihood ratio	19.274	4	.001
No. of valid cases	103		

[a] 3 cells (30.0%) have an expected count of less than 5. The minimum expected count is 2.72.

TABLE 3.3
Crosstabulation Showing "Actual" and "Expected" Frequency between Two Variables "Experience" and "Knowledge about IoT"

Experience and Knowledge about IoT Crosstabulation

			Knowledge About IoT					Total
			1	2	3	4	5	
Experience	0_5	Count	2	0	3	13	1	19
		Expected count	3.7	2.6	1.3	9.8	1.7	19.0
	10_15	Count	9	6	0	8	1	24
		Expected count	4.7	3.3	1.6	12.3	2.1	24.0
	15+	Count	2	1	1	1	0	5
		Expected count	1.0	.7	.3	2.6	.4	5.0
	5_10	Count	7	7	3	31	7	55
		Expected count	10.7	7.5	3.7	28.3	4.8	55.0
Total		Count	20	14	7	53	9	103
		Expected count	20.0	14.0	7.0	53.0	9.0	103.0

The Chi-Square test was run on SPSS and the output obtained is discussed below.

The first table is basically a cross tabulation obtained for the two variables under consideration and consists of both "Expected" and "Actual" frequency count for the cross tabulation. The same will be used to carry out the Chi-Square test. The following table is the output of the Chi-Square test specifying the significance level. Also, the notation of Experience: "0_5" indicates that the number of years of work experience of the respondent surveyed was between 0 and 5 years. Similarly, if the Experience is "10_15," that means the total years of work experience possessed by the respondent lies in the range 10–15 years. Similar understanding can be made for subsequent intervals (Table 3.3 and Table 3.4).

Interpretation: As the likelihood ratio is 0.009, which happens to be less than .05, we reject the null hypothesis and accept the alternative hypothesis (H_1: There is significant relationship between knowledge of IoT and experience).

TABLE 3.4
Chi-Square Test Output

Chi-Square Tests

	Value	df	Asymptotic Significance (2-sided)
Pearson Chi-Square	23.645[a]	12	.023
Likelihood ratio	26.426	12	.009
No. of valid cases	103		

[a.] 15 cells (75.0%) have an expected count of less than 5. The minimum expected count is .34.

Now, we move on to test another hypothesis related to our case study.

Hypothesis 3:

H0: There is no significant association between knowledge of IoT and company size.
H1: There is a significant association between knowledge of IoT and company size.

The Chi-Square test was run on SPSS and the output obtained is discussed below.

The first table is basically a cross tabulation obtained for the two variables under consideration and consists of both "Expected" and "Actual" frequency count for the cross tabulation. The same will be used to carry out the Chi-Square test. The following table is the output of the Chi-Square test specifying the significance level. Also, the notation of company size: "0_50" indicates the number of employees working in the organization surveyed. Similarly, if the size of company is "50_100," that means the total number of working employees in that organization would be somewhere in the range 50–100. Similar understanding can be made for subsequent intervals (Table 3.5 and Table 3.6).

Interpretation: As the likelihood ratio is 0.105, which is greater than 0.05, we fail to reject the null hypothesis, and thus the null hypothesis is accepted (H0: There is no significant association between knowledge of IoT and company size).

Hypothesis 4:

H0: There is no significant association between knowledge of IoT and serving customer.
H1: There is a significant association between knowledge of IoT and serving customer.

The Chi-Square test was run on SPSS and the output obtained has been discussed below-

The first table is basically a cross tabulation obtained for the two variables under consideration and consists of both "Expected" and "Actual" frequency count for the

TABLE 3.5
Crosstabulation Showing "Actual" and "Expected" Frequency between Two Variables "Company Size" and "Knowledge about IoT"

Company Size and Knowledge about IoT Crosstabulation

			\multicolumn{5}{c	}{Knowledge About IoT}	
			1	2	3	4	5	Total
Company size	0_50	Count	2	1	2	5	4	14
		Expected count	2.7	1.9	1.0	7.2	1.2	14.0
	50_100	Count	6	5	2	29	1	43
		Expected count	8.3	5.8	2.9	22.1	3.8	43.0
	100_500	Count	10	6	1	14	2	33
		Expected count	6.4	4.5	2.2	17.0	2.9	33.0
	500+	Count	2	2	2	5	2	13
		Expected count	2.5	1.8	.9	6.7	1.1	13.0
Total		Count	20	14	7	53	9	103
		Expected count	20.0	14.0	7.0	53.0	9.0	103.0

TABLE 3.6
Chi-Square Test Output

Chi-Square Tests

	Value	df	Asymptotic Significance (2-sided)
Pearson Chi-Square	20.626[a]	12	.056
Likelihood ratio	18.377	12	.105
No. of valid cases	103		

[a] 13 cells (65.0%) have an expected count of less than 5. The minimum expected count is .88.

cross tabulation. The same will be used to carry out the Chi-Square test. The following table is the output of the Chi-Square test specifying the significance level. Also, the notation of "Customer Whom You Serve" is basically referring to the type of clients handled by the respondents. The various categories of these respondents are: "Biotech Firms," "Healthcare content and Services," "Hospital and Healthcare Facilities," "Medical Devices Companies," and "Payers and Insurance Companies" (Table 3.7 and Table 3.8).

Interpretation: As likelihood ratio is 0.002, which happens to be less than 0.05, we reject the null hypothesis and accept the alternative hypothesis (H1: There is a significant association between knowledge of IoT and serving customer).

With all four hypotheses tested above, this paper now moves on to discuss the difficulties associated with the IoT infrastructure, the future scope of IoT, and the conclusions that can be drawn overall.

TABLE 3.7
Crosstabulation Showing "Actual" and "Expected" Frequency between Two Variables "Client" and "Knowledge about IoT"

Customer Whom You Serve and Knowledge about IoT Crosstabulation

			\multicolumn{5}{c}{Knowledge About IOT}	
			1	2	3	4	5	Total
Customer whom you serve	Biotech firms	Count	5	0	0	6	0	11
		Expected count	2.1	1.5	.7	5.7	1.0	11.0
	Healthcare content and services	Count	1	1	0	1	0	3
		Expected Count	.6	.4	.2	1.5	.3	3.0
	Hospital and healthcare facilities	Count	6	3	4	30	4	47
		Expected count	9.1	6.4	3.2	24.2	4.1	47.0
	Medical device companies	Count	2	2	3	13	3	23
		Expected count	4.5	3.1	1.6	11.8	2.0	23.0
	Payers and insurance companies	Count	6	8	0	3	2	19
		Expected count	3.7	2.6	1.3	9.8	1.7	19.0
Total		Count	20	14	7	53	9	103
		Expected count	20.0	14.0	7.0	53.0	9.0	103.0

TABLE 3.8
Chi-Square Output

Chi-Square Tests

Value	df
35.915[a]	16
37.501	16
103	

[a] 19 cells (76.0%) have an expected count of less than 5. The minimum expected count is .20.

3.5 DIFFICULTIES FACED BY IOT IN HUMAN SERVICES OR HEALTHCARE

To state that cutting-edge prescription is moving forward would be an understatement. The advancement quickens each day "with no regret," changing all known therapeutic practices. Worldwide social insurance advances are dependent on the most recent accomplishments of the planet's most prominent figures, and include astonishing possibilities for self-governing, self-learning tech arrangements.

Alongside such fast improvement, however, comes an exacting need to stay aware of the pace. Interestingly, all restorative fields are either looking to—or as of now go

Technological Advances in Healthcare

inseparably with—cutting edge innovations, from diagnostics to therapeutics, from pediatrics to complex medical procedures.

Advances are various – man-made brainpower, AI, and so on. Be that as it may, what specific tech idea or blend of ideas can give adequate observing and overseeing forces such an ever growing, worldwide specialty requires? The appropriate response might be found in the Internet of Things or IoT. In spite of the fact that the idea' is relatively new, it has now become firmly associated with medical services, so much so that it is generally called the Internet of Medical Things.

The broad centralization and interconnection benefits that IoT tech gives are hard to overestimate. It allows wellbeing monitoring, remote treatment to an unheard-of level. Yet, we should contemplate IoT controls in social insurance in greater detail. There are people investigating diverse paths with respect to applications of IoT in healthcare, like Real Time Health System, Telemedicine, Aging Setup, etc. This is now being executed around most of the world for the healthcare industry.

There are basically three key models driving all these exercises:

1. There is growing mindfulness among customers about their wellbeing.
2. Diverse healthcare framework players are endeavoring to fulfil people's needs with novel approaches.
3. All the patients need to reduce their healthcare costs through better services.

In many respects the IoT world faces various challenges, including specialized, administrative, market based, and socio-moral considerations [4], but the focus is on ensuring security as this is the vital driver of various problems including assistance from government.

- *Scalability*: As billions of IoT devices get connected to the framework, a tremendous volume of data needs to be organized. The system which stores this information from the IoT gadgets must stay adaptable [4]. The raw information from the world needs deep examination for the comprehension of the crucial information.
- *Interoperability*: Technological models that need to be consolidated are so far isolated in most areas. This will establish the vital IoT contraptions framework. This lack of interoperability balances the environment towards the vision of associated interoperable objects [4].
- *Absence of government support*: Organizations like the FDA should join and have a working impact by setting IoT device regulations for devices and people security [4].
- *Wellbeing of patients*: Whenever IoT devices are used on patients, in view of the nature of IoT devices, any breaches in security are considered incredibly serious.
- *Security and individual assurance*: Security reviews were not correctly conducted [4]. The IoT should ensure the classification and availability of patient sensors in medical facilities. Meanwhile, staff of medical facilities need to reflect more on multiple compliances and guarantee that more

stringent guidelines will be enforced, such as the EU GDPR, where genetic and other health data as well as normal data are considered.
- *B2B versus B2C healthcare IoT*: Most healthcare IoT organizations focus on enhancing quality and encouraging cooperation between client/patient suppliers and payers. In an inexorably industrial Internet of Things, suppliers of human facilities bring about advances through the application of electronic thinking and Big Data. This framework isn't developing as fast as the case for B2C IoT.
- *Designing of difficulties*: Because IoT penetrates at a quicker rate, problems with the framework can be encountered very soon [4]. But difficulties remain today with an IoT-based scheme, such as:
  - Limited essentials of the system
  - Limited memory storage
  - Limited process control system

Reasons for not adopting the technology (IoT-based medical wearables) [13, 14]:

- Traditional barriers
- No intention to adopt the devices
- Usage barriers
- Value of openness to change
- Cost
- Perceived privacy risk

3.5.1 How Practical Is It to Actualize IoT Medical Services Arrangements?

The primary concern of IoT in social insurance is to guarantee more benefit to patients, more engaged human services suppliers, and a cost savings for both. With a solid innovative accessory, you can fabricate a savvy IoT sensor arrangement.

There are sure essentials for achieving cost effectiveness, which can be applied in IoT application advancement for medical services too, for example, diminished overspending, the clearly outlined job of sensors and their abilities, and the sending and receiving of information when vital. For doing this, you have to make a powerful IoT application advancement methodology and engineer it in a joint effort with your innovation accomplice.

3.6 THE FUTURE OF IOT IN HUMAN HEALTHCARE

Below are some things developing in the healthcare sector, which are fundamentally driven by the capacity to screen and control remote devices [5].

- *Ingestible sensor*: This is the progress based on the sensor that is made entirely of fixations discovered in support and incited upon ingestion. After swallowing the pill with the ingestible sensor, a response from the stomach fluid compound gives the necessary energy source. This sensor stays in the stomach of the patient and provides data on the response of patients

to medication in real time [5]. This can be particularly helpful for patients who take medication regularly. This information is also transmitted to the necessary providers of social insurance.
- *Digital drug*: This is a progression of the ingestible sensor. Here, as opposed to an extra pill, the medication itself goes electronic. Each pill includes a tiny sensor that can provide significant information on the body's responsiveness through a computerized well-being report system. This helps to give greater clarity about patient recovery [5]. Advanced medicines are an ongoing project and are not yet cleared by the FDA. Preliminary clinical trials in these restorative areas are currently being conducted.
  - Cardiac failure
  - Transplantation and dialysis
  - Focal nervous system
- *Personalized medication*: The transmission of human services henceforth relies on data from the population. Patients are separated into cohorts characterized by similar side effects in various respects for the majority[5]. These cohorts, subsequently treated with drugs, can include many people. Healthcare industry and electronic health records will drive a tailored way to handle the administration of medication; taking into account patient DNA examination, tailored prescriptions will also enable customization of medicines and effective medication mixes depending on the hereditary makeup of the individual.
- *Predictive analysis*: Using predictive analysis in human services will benefit from the convergence of different data vaults, which have specific patterns and lifestyle propensities. The more information we have from an individual, the more recommendations the analysis will have. Depending on the information focus, the models can be further modified, which eventually results in more precision and success [5]. Sooner rather than later, the wellness pal portable application may warn you days in advance by figuring out the signals in the circulatory system and sending you to your cardiac expert, by which time you will be showing signs of a heart attack.

3.7 CONCLUSION

The healthcare industry is in crisis. Human services administrations are costlier than ever before, the global population is ageing, and the number of diseases is on the rise. What we are approaching is a world where fundamental medicinal services are out of reach of many, an enormous section of society is incapacitated because of old age, and individuals would be progressively vulnerable to ongoing disease. Is it the apocalypse we suspected? In any case, IoT application development might be our salvation.

Where IoT medical devices are headed next is hard to predict—but we are very sure that advantageous things will undoubtedly occur with the development of IoT and finances invested in the healthcare sector. In this chapter we addressed that with machine-to-machine and human-to-machine interfaces, all the physical products will function impeccably. The elements around the sources of genomic information will enable the person's vulnerabilities to be perceived. It is not suggested to replace skilled clinicians with these versatile expert apps. In this methodology of data sources growth, new IoT instances can alter the way fundamental human medical services are delivered to patients. IoT brings new momentum for human services along with extraordinary quality. The recommendation for IoT human services devices are remotely advising, handheld logical devices for recognizing indicators for conditions like intestinal ailments and many other diseases. If you are searching for opportunities in this area, healthcare is a remarkable place to be. We are eager to perceive what sort of developments—from network, to data protection, to application architecture—come into that space.

REFERENCES

1. Gubbi, Jayavardhana, Buyya, Rajkumar, Marusic, Slaven, and Palaniswami, Marimuthu. (2013). Internet of Things (IoT): A Vision, Architectural Elements, and Future Directions. *Future Generation Computer Systems*, 29(7), 1645–1660.
2. Nguyen, H. H., Mirza, F., Naeem, M. A., and Nguyen, M. (2017). A Review on IoT Healthcare Monitoring:Applications and a Vision for Transforming Sensor Data into Real-Time Clinical Feedback. *IEEE 21st International Conference on Computer Supported Cooperative Work in Design*, Milan.
3. Dhananjay, Singh, Gaurav, Tripathi, and Jara Antonio, J. (2014). A Survey of Internet-Of-Things: Future Vision, Architecture, Challenges and Services. *IEEE World Forum on Internet of Things (WF-IoT)*, Seoul.
4. Ndubuaku, Maryleen, and Okereafor, David. (2015). Internet of Things for Africa: Challenges and Opportunities. *International Conference On Cyberspace Governance – Cyberabuja 2015*, Abuja.
5. Omanović-Mikličanin, Enisa, Maksimović, Mirjana, and Vujović, Vladimir. (2015). The Future of Healthcare: Nanomedicine and Internet of Nano Things. *Folia Medica – Facultatis Medicinae Universitatis Saraeviensis*, 50(1), 23–28.
6. Sharma, Cheena, and Sunanda, Dr. (2017). Survey on Smart Healthcare: An Application of IoT. *International Journal on Emerging Technologies* (Special Issue NCETST-2017) 8(1), 330–333.
7. https://www.bloomberg.com/quicktakes/internet-things.
8. https://www.bloomberg.com/professional/blog/IoT-market-predicted-double-2021-reaching-520-billion/.
9. https://www.forbes.com/sites/bernardmarr/2019/02/04/5-internet-of-things-trends-everyone-should-know-about/#30da82ec4b1f.
10. https://www.forbes.com/sites/danielnewman/2019/02/19/IoT-in-2019-what-can-we-expect/#64baf8fbc39a.
11. Haghi, M., Thurow, K., and Stoll, R. (2017). Wearable devices in medical internet of things: scientific research and commercially available devices. *Healthcare Informatics Research*, 23(1), 4. Doi: 10.4258/hir.2017.23.1.4
12. IDC (2017). IDC Forecasts Worldwide Shipments of Wearables to Surpass 200 Million in 2019, Driven by Strong Smartwatch Growth. Available at: www.idc.com/getdoc.jsp?containerId=prUS40846515 (accessed 6 September 2016).
13. Sivathanu, Brijesh (2018). Adoption of Internet of Things (IOT) Based Wearables for Elderly Healthcare – A Behavioural Reasoning Theory (BRT) Approach. *Journal of Enabling Technologies*, 4–12.
14. Karahoca A., Karahoca, D, and Aksöz M., (2018). Examining intention to adopt to internet of things in healthcare technology products. *Kybernetes*, 47(4), 742–770. doi. 10.1108/k-02-2017-0045

4 The Internet of Things (IoT) and Contactless Payments
An Empirical Analysis of the Healthcare Industry

Pooja Ahuja and Sandhya Makkar

CONTENTS

4.1 What Are IoT Payments? 62
    4.1.1 Advantages of IoT in Healthcare 62
    4.1.2 Advantages of Digital Payment for Healthcare Services 63
4.2 Literature Review 63
    4.2.1 Technology Acceptance Model (TAM) 63
        4.2.1.1 Perceived Convenience 65
        4.2.1.2 Perceived Ease of Use 65
        4.2.1.3 Conceptual Framework and Propositions 66
4.3 Proposed System 66
    4.3.1 Proposed Architecture 66
    4.3.2 Patient Position Device 66
        4.3.2.1 Sensors to Measure Body Temperature 66
        4.3.2.2 Sensors to Measure Heart Rate 67
    4.3.3 Recommended Design 67
        4.3.3.1 Implementation 67
        4.3.3.2 Importance in the Healthcare Sector 68
        4.3.3.3 Instamed and Its Digital Pockets Offering 68
        4.3.3.4 Med Rec and Its Offering 68
        4.3.3.5 Process Changes 69
    4.3.4 Methodology 69
        4.3.4.1 Research Design 69
        4.3.4.2 Questionnaire Design 69
    4.3.5 Hypothesis 69
    4.3.6 Data Analysis and Design 70

4.4 Results 72
4.5 Discussion and Conclusion 72
4.6 Limitations 72
4.7 Conclusion 73
4.8 Findings 73
Bibliography 73

4.1 WHAT ARE IOT PAYMENTS?

The healthcare industry is about to be transformed through the use of digital technologies. This will be seen by looking at how hospitals will operate their banking transactions and payment process. A team at one of the leading banks has shown that how the healthcare sector can control technology to improve its banking operations in order to become more efficient as well as improve outgoing patients' hospital experience. They further defined how the bank has cooperated closely with many players in the healthcare ecosystem, like third-party administrators (TPAs), insurance companies, hospitals, clinics and third-party technology companies in order to determine the concerns of the healthcare industry, and has then formed a value scheme that can help the healthcare industry.

The team also elaborated that the present government is promoting digitization across industries through policy, incentives, and investments. With regards to the banking sector, they mentioned that

> 100 crore Aadhar cards have been issued in India which are now been linked to bank accounts. The Bharat Interface for Money (BHIM) Unified Payments Interface (UPI) App was recently launched by the government that lets users make payments using Unified Payment Interface. This shows that the government is lending its full support in its digital India campaign.

According to the team, hospitals must go down the digital route to become future ready. Nowadays, patients are looking for faster, inexpensive, and hassle-free ways to pay; it is not easy for this sector to adjust and bring itself into line with current technology to be able to fulfill patients' needs. Various banks are looking for partnership prospects in the healthcare sector and are discovering many areas wherein banks can be an active partner in the current ecosystem. There are various technical aspects involved in choosing digital banking solutions for hospital bill payment. The proposed banking solutions can add value to the patient experience as well as support operational competences. The Unified Payment Interface (UPI) payment system for the healthcare sector will simplify the patient payment process and has the ability to reduce patient wait times we well as human errors in payments processing. Hospitals can have reconciliation solution for payments made by TPAs and insurance companies, UPI solutions, and cash deposit machines, to name a few.

4.1.1 Advantages of IoT in Healthcare

- IoT plays a critical role in the enhancement of regular healthcare systems, and its features will fulfill a number of patients' well-known needs.

IoT and Contactless Payments 63

- The improvement of equipment in the healthcare sector using IoT allows game-changing and life-enhancing offerings throughout the economy.
- The use of IoT in the healthcare sector has created many opportunities and issues to be addressed.
- According to a forecast, the IoT healthcare market is worth $163.57 billion and will involve with 53 billion connected gadgets by 2020.

4.1.2 Advantages of Digital Payment for Healthcare Services

- Digitalization can help bring down costs and will lead to better care, with improved access to data and coordination among various medical professionals.
- Digital billing can help reduce healthcare fraud and errors. Sending bills and reminders of payments through email and text can result in fewer frauds.
- Linking phone numbers with hospital records can facilitate live interaction between patients and doctors.
- Patients can set up a payment plan, and can have clear terms and medical practice branding so that they do not get confused about what they are paying for or whom they are paying, which are known to be common obstacles to prompt payment of medical bills, according to experts.

4.2 LITERATURE REVIEW

The relationship between the concept of IoT and its role in healthcare applications is examined here, and devices used for implementing IoT technologies are introduced. This paper focuses on the difficulties faced by patients and the solutions offered by various devices in monitoring and tracking diseases through digital medicine.

Advanced technologies in IoT such us cloud computing, Big data, grid computing, and soft computing can be used in handling connected devices for monitoring and guiding patient health. Blood-pressure and body-temperature monitoring, electrocardiogram monitoring, wheel-chair management, glucose-level sensing, and emergency healthcare can be provided through connected devices anywhere in the world.

From the reviews, devices like Warfarin, wearables, automated wheel chairs, wireless transmitters and receivers, and activation kits can be emphasized. This paper examines the IoT technologies used for remote healthcare monitoring and looks at the challenges and obstacles faced by IoT.

4.2.1 Technology Acceptance Model (TAM)

TAM is used as the theoretical support in this study. It represents a significant theoretical support towards understanding IoT usage and the reception behavior of patients. The aim of TAM is "to provide an explanation of the determinants of computer acceptance that is general, capable of explaining user behavior across a broad range of end user dividing skills and user populations, while at the same time being both parsimonious and theoretically justified". It tries to forecast and demonstrate system usage by suggesting that two theories, *perceived usefulness* and *perceived ease of use*, are the most significant in determining consumer acceptance behavior. TAM explained the concepts of perceived usefulness as the amount to which a consumer thinks that if he would use a specific system by which his job performance would be enhanced, and perceived ease of use as the amount to which a consumer thinks that if he would use a specific system, he would not have to make many efforts.

Some earlier studies which used TMA have shown that perceived usefulness is very important element in the willingness to use a technology, while perceived ease of use is very important when devices are being used continually.

User reception and assurance are critical for the expansion of any new technology. Moreover, *reception* has been observed as a function of user participation in systems improvement. Generally, *reception* means "an antagonism to the term rejection and means the encouraging decision to use a revolution". Decision makers are interested to know the issues that impact on users' decision to use a particular system so that they can take those into consideration when improving the technology. Both practitioners and researchers need to know why people agree on new technologies. The answer might help them to come up with better methods of designing, evaluating, and predicting the responses of users in using new technologies. Various theories including TAM have been beneficial in a broad range of domains to understand and predict users' behavior such as voting, dieting, family planning, donating blood, women's occupational orientations and breast-cancer examination, choice of transport mode, turnover, use of birth-control pills, education, consumer purchasing behaviors, and computer usage. Some investigating the field of TAM have developed frameworks to measure the usage of particular developed and instigated technology.

Numerous models and structures have been established to describe user acceptance of new technologies, and these models show factors that can affect user reception such as TAM, the theory of scheduled behavior and diffusion of innovation, the theory of reasoned action, the model of PC utilization, the motivational model, the integrated theory of reception and use of technology and social perceptive theory. Many studies have used these traditional frameworks to conduct their research, and the rest combined previous models or added new constructs to develop models to carry out their study.

More than one hypothetical approach is necessary for a complete understanding of the issues involved, and for clarity, approaches are treated unconventionally. However, engaging in thorough theoretical consideration of the issues involved requires established methods. Therefore, an overview of the available overall implementation models are essential in this field. In this chapter, implementation theories and models are presented to give a better understanding of these models and theories.

Though perceived convenience is widely recognized as the major factor affecting usage, it is fundamentally appropriate to explain in more detail the fundamentals of this factor (see Figure 4.1).

According to the study by Abbasi et al. (2013), Technology Acceptance Model (TAM) has been used in various research projects around the world and in different fields of study, and it may be duplicated or extended to suit the framework of a study. It is for this reason that this analysis paper proposes the use of TAM with some very minor modifications (removal of "attitude toward using" and "actual system use"),

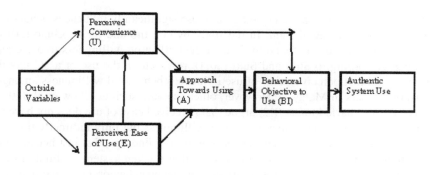

FIGURE 4.1 Technology Acceptance Model (TAM). Source: Davis (1989).

considering the comparative result of the freelance variables (perceived utility, perceived easy use) on the constant purpose of using an electronic collection system in hospitals.

4.2.1.1 Perceived Convenience

According to Rouibah and Abbas (2006), the more the number of users of a system area unit, the more the output and potency in their work can be increased through the utilization of the system. Yadav et al. (2010) contend that users' intentions are regulated by their perception of the quality of the system. Various studies have found that perceived quality has a very important and robust influence on individuals' behavioral intention to use a brand-new technology or system. According to a study by Sambasivan et al. (2010) on e-procurement in developing countries, perceived utility has important influence on the active intention to use an e-procurement system within public-sector establishments. Tella and Olasina (2014) found in another study that the activity intention among bank customers to use electronic payment was influenced by its perceived utility in banking transactions. Diatmika et al. (2016) revealed that perceived quality could be an issue that influences behavioral objectives in most of the organizations that have adopted an accounting data system (AIS). Therefore, perceived quality is taken into account for its vital influence on behavior when it comes to using digital payments in hospitals.

The following proposition can be made:

Proposition 1: Perceived usefulness absolutely encourages the behavioral objective to use digital payment system in hospitals.

4.2.1.2 Perceived Ease of Use

According to Davis (1989), *perceived ease of use* can be defined as the range to which a system user considers that the use of that system might make him free from making efforts. That means the user has seen that the system is very simple to use. Therefore, it is believed that users' behavioral purpose increases when the system is not very difficult to use. A study by Terzis and Economides (2011), was conducted on the use of a computer-based assessment (CBA) system amongst undergraduate students in a college, which depicts that perceived ease of use significantly influences behavioral purpose to use any new system. In the study by Guritno and Siringoringo (2013), perceived ease of use was found to have a direct effect on willingness to use e-airticketing. According to Suki and Suki (2011), research on the use of a 3G mobile service found that perceived ease of use influences behavioral willingness amongst smartphone users in Malaysia. However, other scholars such as Chow et al. (2012) and Nasri and Charfeddine (2012) argue that perceived ease of use does not have an uninterrupted significant influence on user willingness but only indirectly through perceived usefulness. That means, according to their findings, that just because the perception of an information system's user is that a given system is simple to use does not guarantee the direct probability of its effect on behavioral aim to use, unless if the user is responsive to its usefulness. This study recommends determining the significant influence of perceived ease of use on behavioral willingness to use an electronic collection system in various hospitals based on the following proposition:

Proposition 2: Perceived ease of use evidently encourages the behavioral objective to use an electronic collection system in hospitals.

4.2.1.3 Conceptual Framework and Propositions

It can be seen that perceived usefulness and perceived ease of use were used in various past studies to determine the behavioral objective of system or technology use among individuals. In view of that, the theoretical structure of this study was developed from the Technology Acceptance Model (TAM).

4.3 PROPOSED SYSTEM

4.3.1 Proposed Architecture

The proposed structure consists of a range of utilities such as administrative application, health practitioner application, android utility, and cloud server. A hardware system is connected to the android smartphone via Bluetooth. In this paper, we will study the use of three sensors: a heart rate sensor, a body temperature sensor, and a physical position sensor.

4.3.2 Patient Position Device

The Patient Position device, commonly known as Accelerometer, records five distinct patient positions (prone, supine, standing/sitting, left, and right.) In many cases, it is imperative to screen the physical positions and movements in terms of their relationships to specific sicknesses (i.e., sleep disorders and syndromes). The bodyoperated device should also detect fainting or falling of elderly people or persons with disabilities and understand if someone stops standing via a movement detector.

4.3.2.1 Sensors to Measure Body Temperature

The normal core body temperature taken internally is 37.0°C, i.e., 98.6°F. For adults, body temperature by varies about 0.5°C, or 0.9°F, during the daytime, with temperatures

IoT and Contactless Payments

being lower in the morning and higher in late afternoon and evening; the change in temperature is dependent on the body's requirements and activities. Timely measurement of body temperature is very important. The objective of the same is that by changes in body temperature can be attributed to various different diseases. Similarly, through measuring body temperature, the course of such ailments can be monitored, and the efficiency of any treatment can be evaluated with the aid of the physician.

4.3.2.2 Sensors to Measure Heart Rate

A person's normal heart rate can increase and decrease; variations between 50 and 60 beats per minute (bpm) are in the healthy range and do not require unusual attention. The term *tachycardia* is defined as a resting heart rate above 100 bpm, despite a resting state ranging between 80–100 bpm in general. If a person displays elevated heart rate during sleep, it may additionally be a symptom of hyperthyroidism or anemia. The wide variety of sensors can be multiplied up to eight. The hardware system consists of several components: a Bluetooth controller, a microcontroller, an ADC (analog to digital) converter, a signal conditioner, and a buzzer. The sensors are connected to the signal conditioner, through which all the values from the sensors are collected and then sent to the ADC. There, the signal is transformed from analog to digital and is then sent to the microcontroller which, transmits the data to the android software via a Bluetooth controller. In case of emergency, the buzzer will ring or an SMS/email will be dispatched to the health practitioner via their smartphone. All the data related to the patient are stored on the cloud server. We have brought cloud servers to the point where consumers of exclusive platforms can be interfaced with them, which overcomes the downside of using servlets.

The patient can be observed for signs of unconsciousness or death from the phone camera. The status of the patient can also be recorded and dispatched after a fixed time interval. Also in an emergency situation, the beeper would blink and a display will be sent via SMS or email.

Near field communication (NFC) tags can be linked with the help of an NFC-enabled smartphone cab that can be used by a medical doctor to recover the patient's information. The use of NFC tags eliminates the paperwork that had to be finished throughout the process of registration of the patient when he or she is admitted.

4.3.3 RECOMMENDED DESIGN

4.3.3.1 Implementation

A special identity is allotted to every patient when he or she is admitted to hospital for the first time in the form of NFC tag, which is linked with NFC-enabled smartphones and given to doctors and other staff. An NFC care application is used to check the content material coming from the phones to the NFC tag. The administrator can create a special tag ID and software link with the aid of the use of the NFC problem application. To create any patient's application, the hyperlink administrator uses the IP of the server and then switches to cell by means of the same IP.

Whenever the NFC tag is situated near to NFC-enabled smartphones, the patient information is immediately saved to the server. Doctors similarly access the server to get the patient's data.

After the patient is admitted to hospital, the smartphone linked with the NFC would have signal exposure to be updated on the status of that patient. Moreover, different sensors can reveal separate factors such as heart rate, body temperature, etc.

4.3.3.2 Importance in the Healthcare Sector
- Electronic options are already altering the longstanding methods that are typical of fitness care.
- In case of digital pockets that is customer friendly, it can help grant alternatives to patients and also reduce the paperwork that is associated with healthcare payments.
- This technology can have a real impact on a person's experience of a healthcare institution.
- Studies show that patients who are satisfied with their billing experience in a hospital are more likely to recommend that hospital to others.

4.3.3.3 Instamed and Its Digital Pockets Offering
- This solution provider offers a repayments network for healthcare providers. They can help connect consumers or payers as nicely as carriers on a single and impassable platform.
- InstaMed (Healthcare and Payment Solutions) showcases countless charge vendors on the digital and impenetrable platform that can assist to minimize the dangers or complexities of digital repayments for a healthcare institution.
- This platform presents charge as well as clearinghouse solutions. Providers will be in a position to acquire money efficiently, and payers can limit their time spent for series as well as prices.
- Settlement and postage prices for claim payments can be decreased through connecting all payers of a healthcare organization in the Instamed network.
- Payers can customize their payment options and pay premiums or make payments to a healthcare provider using particular kinds of payment accounts.
- This solution provider provides a secure and seamless healthcare price framework. They have a presence in fifty states across the United States and are delivering outcomes for them in payment transactions at a digital level.

4.3.3.4 Med Rec and Its Offering
This is a Blockchain-based totally digital scientific data administration gadget that can help a healthcare institution to manage individuals' electronic statistics and transaction records.

As many Blockchain-based options assist in creating a more straightforward way to get right of entry to scientific archives and be able to edit them, that is also something that MedRec offers.

Patients' documents can be linked and accessed throughout several database systems of healthcare providers. The machine will act as an interface between the healthcare archives of individual institutions as well as ensuring that private data from health devices are up to date on the money owed by users.

Patients will be able to grant medical doctors access to their private and health information. This can also easily be forwarded to family members as well as healthcare providers or researchers.

The system would allow control of and access to personal health data users. It would also have an electronically generated bill for the individual transactions completed on the platform.

4.3.3.5 Process Changes

An e-wallet consists of information about price plans, automatic repayments, and maximum payment limits. Digital wallets also record communication preferences, for example, whether or not a patient is still enrolled for e-statements or communications via text message.

The technology of digital wallets has begun to convert transactions from cash to digital payments. Hospital staff used to send paper statements in the mail; nowadays they are sending e-statements. They used to spend time following up payments over the phone or by e-mail, apprising patients about payment modes; now staff can send important notifications via text message and can send guidelines automatically so that the patient is not confused about how to make payments for hospital services.

4.3.4 Methodology

4.3.4.1 Research Design

This section explains the strategy and processes used for the study. For this study, a survey methodology was developed and the survey was administered to hospital outpatients in New Delhi city.

A stratified sampling technique was utilized to support the categories, and two teams of different age groups were compiled from every stratum, creating a complete survey of 104 respondents. The unit of study is individual in order to make sure equal opportunity is maintained. Therefore, a total of 104 questionnaires were arbitrarily distributed to the targeted respondents.

4.3.4.2 Questionnaire Design

As mentioned above, a survey form was employed for this study that consisted of the proper items for measuring the study variables. Also, in compliance with the scientific norms of modern research, all the measurements were changed from previous studies associated with the study. For instance, perceived usefulness (PU), perceived simple use (PSU) and behavioral intention (BI) were all changed from the original technology acceptance model (TAM), with some minor modifications to match the framework of digital payment use in hospitals. In this study, two factors were employed in the mensuration of the study variables: the age of respondents and the mode of payment employed by them.

4.3.5 Hypothesis

- The null hypothesis is H0, which means the age of respondents is not associated with mode of payment.

- The alternative hypothesis is H1 which means the age of respondents is associated with mode of payment (see Tables 4.1, 4.2, 4.3, 4.4, and 4.5).

4.3.6 Data Analysis and Design

Data from 104 respondents were collected. They were classified in terms of age (old and young) and in terms of mode of payment in hospitals (cash, cashless, or both). Age and mode of payment preferred are categorical data.

TABLE 4.1
Emergence of Pharmaceutical Industry Growth with Industrial IoT Approach

S. No.	Age	Mode of Payment
1	Young	Both
2	Young	Cashless/digital
3	Young	Cashless/digital
4	Young	Cashless/digital
5	Young	Both
6	Young	Cashless/digital
7	Old	Cashless/digital
8	Young	Cashless/digital
9	Young	Cashless/digital
10	Young	Cash
11	Young	Cashless/Digital
12	Young	Cashless/Digital
13	Old	Cashless/Digital
14	Young	Cashless/Digital
15	Young	Cash
16	Young	Cashless/Digital
17	Young	Cashless/Digital
18	Old	Cashless/Digital
19	Young	Cash
20	Young	Cashless/Digital
21	Young	Cashless/Digital
22	Old	Cashless/Digital
23	Young	Cashless/Digital
24	Young	Both
25	Old	Cashless/Digital
26	Young	Cashless/Digital
27	Young	Both
28	Young	Both
29	Young	Both

TABLE 4.2
Case Processing Summary

	Cases					
	Valid		Missing		Total	
	N	Percent	N	Percent	N	Percent
VAR00003 * VAR00010	104	100.0%	0	0.0%	104	100.0%

TABLE 4.3
Cross Tabulation

			VAR00010			
			Both	Cash	Cashless/Digital	Total
VAR00003	Old	Count	13	2	24	39
		Expected count	11.6	3.8	23.6	39.0
		% within VAR00003	33.3%	5.1%	61.5%	100.0%
		% within VAR00010	41.9%	20.0%	38.1%	37.5%
		% of total	12.5%	1.9%	23.1%	37.5%
	Young	Count	18	8	39	65
		Expected count	19.4	6.3	39.4	65.0
		% within VAR00003	27.7%	12.3%	60.0%	100.0%
		% within VAR00010	58.1%	80.0%	61.9%	62.5%
		% of Total	17.3%	7.7%	37.5%	62.5%
Total		Count	31	10	63	104
		Expected Count	31.0	10.0	63.0	104.0
		% within VAR00003	29.8%	9.6%	60.6%	100.0%
		% within VAR00010	100.0%	100.0%	100.0%	100.0%
		% of total	29.8%	9.6%	60.6%	100.0%

TABLE 4.4
Chi-Square Tests

	Value	Df	Asymptotic Significance (2-sided)
Pearson Chi-Square	1.576[a]	2	.455
Likelihood Ratio	1.701	2	.427
No. of Valid Cases	104		

[a] 1 cells (16.7%) have expected count less than 5. The minimum expected count is 3.75.

TABLE 4.5
Symmetric Measures

		Value	Approximate Significance
Nominal by Nominal	Phi	.123	.455
	Cramer's V	.123	.455
No. of Valid Cases		104	

4.4 RESULTS

Results of the Chi-squared: From the top row of the output table, it can be seen that the Pearson Chi-Square statistic is $\chi^2 = 1.576$, degrees of freedom 2, corresponding to $p < 0.001$. Therefore, the null hypothesis is rejected with 99.9% confidence, and it has been concluded that there is very strong evidence of an association between age and mode of payment.

From the top row of the last table, the Pearson Chi-Square statistic is $\chi^2 = 1.576$ and $p < 0.455$, which means there is much less chance of the observed data being under the null hypothesis of no relationship.

The null hypothesis is rejected due to $p < 0.05$.

4.5 DISCUSSION AND CONCLUSION

The main objective of this study was to calculate the amount of relation of patients who would prefer to use digital payment to pay hospital bills for varied services in a city hospital. As such, two hypotheses were proposed to look at the numerous relationships between age and mode of payment used for paying hospital bills associated with perceived utility, perceived stress-free use, and perceived self-efficacy on the social objective to use this system among all hospital staff.

4.6 LIMITATIONS

The healthcare sector's antiquated billing process negatively impacts provider revenue on two fronts:

- Administrative tasks like insurance coverage verification, sending and receiving bills, and obtaining prior authorization are costing healthcare providers $9.5 billion annually—which could be saved by automating claims processing and billing patients.
- And a lack of billing options at the point of care could undermine patient loyalty. More than half (57%) of adults in the United States say they have received a surprise medical bill, and 82% say hospitals are "very" or "somewhat" responsible for their surprise bills, according to new research from NORC at the University of Chicago.

4.7 CONCLUSION

This paper proposed a research framework to examine the influence of perceived utility and perceived simple use on the frequency of the use of digital payment wallets by hospitals to discharge its bills. This study may enhance the present body of data by adopting variables in a whole new kind of system (e-collection system). Moreover, the study is planned to specialize in verifying the proposed framework in an exceedingly structured atmosphere, wherever system use is in place either at a preliminary stage or required to use. Hence, this paper may contribute to system studies by giving an initial understanding of and data on the influence of perceived utility and perceived simple use on the behavioral intention of staff towards technology use in the context of government obligation. This paper may become a guide for researchers to conduct more empirical research on the relationships explained in the abstract model. It is strongly believed that the empirical findings of such a survey could greatly assist in gaining insight into the attitudes of public health workers toward the employment of the latest technology in different regions, and particularly the use of electronic systems.

4.8 FINDINGS

- Implementing new digital payment tools could help providers make the most of revenue and increase patient satisfaction. Incorporating online portals with automated claims processing and SMS billing reminders could alleviate some of the administration costs associated with patient billing.
- Allowing patients to prepay online or use digital wallets at the point of care could offer a more consumer-friendly billing process and reduce surprise medical bills. And patients are open to these solutions: According to a study, around 61% of the customers expressed an interest in using digital wallets to make a healthcare payment.
- The current inefficiency of the billing process creates an immense market opportunity for new and existing healthcare payments players.
- Payment processing company Square is expanding its partnerships with healthcare providers and is considering developing more complex healthcare payment systems that could integrate with electronic health records (EHRs). Both developments point to EHRs playing a central role in the evolving $3.3 trillion healthcare spending market.

BIBLIOGRAPHY

Abbasi, M.S., Shah, F., Doudpota, S.M. et al. (2013). Theories and models of technology acceptance behavior: A critical review of literature. *Sindh University Research Journal (Science Series)*, 45(1), 163–170.

Ahmed, F., Burt, J. and Roland, M. (2014). Measuring patient experience: Concepts and methods. *Patient*, 7(3), 235–241.

Ajzen, I. and Fishbein, M. (1980) *Understanding Attitudes and Predicting Social Behavior.* Prentice-Hall, Englewood Cliffs.

Akande, L. (2015). Buhari orders federal ministries, agencies to open treasury single account. *Premium Times*. http://www.premiumtimesng.com (23 August 2015, date last accessed).

Alsharayri, M. (2012).Evaluating the performance of accounting information systems in Jordanian private hospitals. *Journal of Social Sciences*, 8(1), 74–78.

Asangansi, I.E., Adejoro, O.O., Farri, O. andMakinde, O. (2008). Computer use among doctors in Africa: Survey of trainees in a Nigerian teaching hospital. *Journal of Health Informatics in Developing Countries*, 2(1), 10–14.

Asogwa, B.E. (2013). Electronic government as a paradigm shift for efficient public services: Opportunities and challenges for Nigerian Government. *Library Hi Tech*, 31(1), 141–159.

Australian Commission on Safety and Quality in Health Care.Core, Common questions for patient experience surveys in hospitals [EB/OL]. http://www.safetyandquality.gov.au/our-work/information-strategy/indicators/hospital-patient-experience/ (20 January 2016, date last accessed).

Bank for International Settlements (1996) *Implication for Central Banks of the Development of Electronic Money*. Basle, ISBN 92-9131-059-X.

Bello, I.S., Arogundade, F.A., Sunusi, A.A. et al. (2004). Knowledge and utilization of information technology among healthcare professionals and students in Ile-Ife, Nigeria: A case study of university teaching hospital. *Journal of Medical Internet Research*, 6(4), e45.

Browne, K., Roseman, D., Shaller, D. et al. (2010). Measuring patient experience as a strategy for improving primary care. *Health Affairs*, 29(5), 921–925.

CAHPS Hospital Survey. HCAHPS survey[EB/OL]. http://www.hcahps.org/surveyinstrument.aspx (18 January 2016, date last accessed).

Crow, R., Gage, H., Hampsom, S. et al. (2002). The measurement of satisfaction with healthcare: Implications for practice from a systematic review of the literature. *Health Technol Asses*, 6(32), 1.

Davis, F.D., Bagozzi, R.P. and Warshaw, P.R. (1989). User acceptance of computer technology: Comparison of two theoretical models. *Management Science*, 35(8), 982–1003. doi:10.1287/mnsc.35.8.982.

Diatmika, I.W.B., Irianto, G., and Baridwan, Z. (2016). Determinants of behavior intention of AIS based information technology acceptance. *Imperial Journal of Interdisciplinary Research*, 2(8), (2016).

Gefen, D., Karahanna, E. and Straub, D.W. (2003). Trust and TAM in online shopping: An integrated model. *MIS Quarterly*, 27(1), 51–90.

Guritno and Siringoringo. (2013) Procedia. *Social and Behavioral Sciences*, 81(June 2013), 212–216.

Hu, L.T. and Bentler, P.M. (1999). Cutoff criteria for fit indexes in covariance structure analysis: Conventional criteria *versus* new alternatives. *Structural Equation Modeling*, 6(1), 1–55. doi:10.1080/10705519909540118.

Jenkison, C., Coulter, A. and Bruster, S. (2002). The picker patient experience questionnaire: Development and validation using data from in-patient surveys in five countries. *International Journal of Quality in Health Care*, 14(5), 353–358.

Labarère, J., Fourny, M., Jean-Phillippe, V. et al. (2004). Refinement and validation of a French in-patient experience questionnaire. *International Journal of Health Care Quality Assurance Incorporating Leadership in Health Services*, 17(1), 17–25.

Mathieson, K. (1991). Predicting user intentions: Comparing the technology acceptance model with the theory of planned behavior. *Information Systems Research*, 2(3), 173–191.

Peter, P.J. and Olson, J.C. (1990) *Consumer Behavior and Marketing Strategy*, 2nd edition. R.R. Donnelley, Chicago.

Sambasivan, M., Patrick, G.W., and Rose, R.C. (2010). User acceptance of a G2B system: A case of electronic procurement system in Malaysia. *Internet Research*, 20(2), 169–187. https://doi.org/10.1108/10662241011032236.

Szajna, B. (1996). Empirical evaluation of the revised technology acceptance model. *Management Science*, 42(1), 85–92. doi:10.1287/mnsc.42.1.85.
The OutPatient Experiences Questionnaire (OPEQ). (2005). Data quality, reliability, and validity in patients attending 52 Norwegian hospitals. *Quality & Safety in Health Care*, 14(6), 433–437.
Tella, A., and Olasina, G. (2014). Predicting users' continuance intention toward e-payment system: An extension of TAM. *International Journal of Information Systems and Social Change*, 5(1), 47–67. https://doi.org/10.4018/ijissc.2014010104.
Terzis, V. and Economides, A.A. (2011). The acceptance and use of computer based assessment. *Computers & Education*, 56(4), 1032–1044. https://doi.org/10.1016/j.compedu.2010.11.017.
Yadav, R., Sharma, S.K., and Tarhini, A. (2010). A multi-analytical approach to understand and predict the mobile commerce adoption. *Journal of Enterprise and Information Management*, 29(2), 222–2.

5 AI in Healthcare

Ghous Bakhsh Narejo

CONTENTS

5.1 Global Research Update on AI in Healthcare 77
5.2 Scope of AI in Healthcare 78
5.3 Data Analytics (Machine Learning, Deep Learning Algorithms) 79
5.4 Medical Imaging (MRI, CT, Ultrasound, and Other Imaging Techniques)....80
5.5 Cancer Detection, Disease Management, and Drug Discovery 81
    5.5.1 AI in Oncology 82
    5.5.2 Detection Using AI 82
    5.5.3 Disease Characterization Using AI 83
    5.5.4 Monitoring of Disease Using AI 83
5.6 Challenges for AI in Healthcare 83
5.7 Future Prospects 84
5.8 Healthcare in 2019 and Beyond: Five AI Trends to Watch 85
    5.8.1 The Internet of Medical Things (IoMT) 85
    5.8.2 Telemedicine 86
    5.8.3 Wearables in Healthcare 86
    5.8.4 Chatbots 86
    5.8.5 Cloud Computing 86
5.9 AI in Healthcare: An Innovation and Opportunity in Disguise 87
5.10 The Story beyond 2019 88
    5.10.1 AI-Powered Chips 89
    5.10.2 IoT and AI 89
    5.10.3 Automated Machine Learning 89
    5.10.4 The Rise of Facial Recognition 89
    5.10.5 Increased Automation 90
5.11 Limitations 90
5.12 Conclusions 91
References 92

5.1 GLOBAL RESEARCH UPDATE ON AI IN HEALTHCARE

Research projects on artificial intelligence (AI) in health and medicine between 1977 and 2018 (84.6% were dated 2008–2018) based upon 27,451 publications reveal a focus upon clinical prediction and treatment through application of major techniques covering research areas such as robotics, machine learning, artificial neural networks, artificial intelligence, and natural language processes. If we classify the research effort on the basis of the health issues examined, cancer-related research represents the highest number of publications, followed by research on heart disease, stroke, vision impairment, Alzheimer's, and depression. These research trends necessitate protocols and regulations to be used when including AI-based products in diagnostics and therapeutics [1].

AI-managed machine learning and deep learning algorithms can be used in a range of healthcare applications, such as helping in medical diagnostics, assisting doctors during consultations, helping in supervising personalized disease treatments, assisting in managing electronic health records, and supporting drug research. Amazon's Alexa assists in diagnosis and prevention of diabetics. Google has come up with the Deep Variant, an AI genomics tool, supported by the cloud that can accurately predict a person's genome [2].

Artificial intelligence comprises two concepts: *artificial*, meaning "un-natural" or "man-made"; and *intelligence*, meaning the "ability to perform tactfully while interacting with the information-rich environment". AI refers to the invention of new intelligent machines which can work and respond as humans when interacting with a new environment. AI is a sub-branch of computer science. Currently, machines, especially computers, are assisting humans and society in activities as diverse as healthcare, warfare, making music, writing books, assessing resumes, checking creditworthiness, and managing the pictures you take on your phone. It can be opined that the machines or computers are making decisions in your life and affecting the people and society around you, whether you like it or not. Recent decades have seen a lot of misuse of the term AI, as it has been employed to enhance marketability and attract customers at the risk of misinformation. On the other hand, a lot of industrial products in the areas of healthcare and home automation are presumed to be assisting humans in their lives. These products and gadgets are mostly operating as computers, employing what is called *machine learning*.

Machine learning is again a sub-branch of computer science, "A science which provides the machines a capability to independently maneuver and act in the complex environments without being programmed in a detailed manner" (https://www.forbes.com/sites/bernardmarr/2018/02/14/the-key-definitions-of-artificial-intelligence-ai-that-explain-its-importance/). Machine learning can be understood as a step toward AI. Machine learning is based upon data-driven intelligent algorithms—the routines and tasks that machines do without explicit programming. Advanced machine learning, also known as deep learning, uses more advanced algorithms to bring machines closer to their goal that is, attaining AI.

5.2 SCOPE OF AI IN HEALTHCARE

If we look at the research trends shown above and the global trends in the emergence of startups or companies that are focused on AI-based healthcare products, we can find some areas of overlap which indicate a general consensus among the research community, those in the medical profession, and entrepreneurs who work on products and solutions. The following are global trends found across the board in all these stakeholders active in the field of AI in healthcare [1–3]:

1. Data analytics (machine learning, deep learning algorithms)
2. Medical imaging (magnetic resonance imaging [MRI], computed topography [CT] scanning, ultrasound, and other imaging techniques)
3. Cancer detection, disease management, drug research

4. Telemedicine, e-health
5. Genetics

Healthcare and AI represent a great combination, affecting four stakeholders: patients, doctors, researchers, and tele-health professionals [2].

AI is set to assist healthcare while copying the traits of human intelligence in the following areas:

1. Assisting doctors and patients in examinations and cataloging of medical data
2. Assisting in designing novel algorithms and techniques for medical decision-making and research
3. Assisting the medical profession in harmonizing the cognitive sciences with medical sciences
4. Assisting in the futuristic data-rich scientific field

The enhanced homogenization of AI techniques with existing medical practices is all set to increase efficiency in treating the patients while reducing the risk of erroneous diagnosis at the same time. It helps to improve pre-operative planning and reduces post-operative complications [1, 2].

5.3 DATA ANALYTICS (MACHINE LEARNING, DEEP LEARNING ALGORITHMS)

The application of AI in performing advanced endeavors including automated routine tasks has evolved mainly into MRI and computed tomography (CT) systems. These systems have additional characteristics to gather the requisite data, and they need to be standardized in order to integrate with tested and traditional database systems.

It must be emphasized here that there are real-world problems which serve as the biggest challenges in the path to adopting AI in the automation of healthcare. This biggest bottleneck is none other than starting capital, which includes the cost of equipment as well as the cost of infrastructure. These newly envisaged sophisticated systems mostly require newly constructed installations as well as skilled staffing, all of which come at very high cost. These and other research-level challenges also hinder the adoption of AI in healthcare.

Cutting-edge research into AI, especially neural networking, natural language processing, image recognition, and speech recognition/synthesis, are contributing to the field, especially its use in healthcare, as all these subfields have yielded highly promising research results on a case-to-case basis.

A typical AI application assisting medical professionals and technicians as well patients and their families can be visualized as an exercise in the automation of a range of routine medicinal tasks as well as other routine tests and precautionary checkups:

1. AI systems can manage medical emergencies, medical alerts, and patient-related reminders. AI can integrate routines which were previously conducted manually, such as machine scans, patient data from lab tests,

the scheduling of medicinal prescriptions, and other alerts for medical professionals.

The alerts and reminders, if set and implemented appropriately, may allow complex AI algorithms to be connected to a display which tracks patient data on vital signs, thereby rapidly assessing and reporting the patient's condition.
2. AI systems can be used to plan diagnostics and therapy for a particular set of symptoms well in advance, allowing detailed plans to be made for the best treatment for a given patient. There are added advantages if the AI systems are incorporated in planning as this automation can design routines based upon the available patient data on specific conditions, adding great value for patients as well as doctors and other medical professionals.
3. Search algorithms have major applications in AI and healthcare as these search algorithms can be utilized for highly complex healthcare applications. Patient data and medicinal as well as other information on medicine and health can be retrieved and upgraded in a highly efficient way using these specialized routine algorithms.
4. Employing image processing techniques and algorithms, many complex medical images can be done very quickly. These may range from an X-ray detecting a simple medical condition to extremely complicated images taken using angiograms, CT scans, and MRI scans. AI systems have been widely tested during image recognition and interpretation for detecting the diseases.

The research component has remained at the forefront of designing expert systems and decision support systems (DSS). The DSS is programmed to have access to big data and the data is manipulated to get specific results. Some examples of DSSs are heart monitoring through expert systems and working towards futuristic electrocardiogram (ECG) systems. Other expert systems being developed concern expert medical imaging systems, clinical laboratory analysis systems, respiratory monitoring systems, electroencephalography systems, and anesthesia systems.

In addition to the above, AI has started impacting surgical robotics by empowering advanced robots automated to undertake surgical tasks at highly efficient success rates. One challenge remains, however, is the replication of human intelligence and body dynamics in AI. Nevertheless, robotics has advanced significantly and has now been initiated in a diverse set of fields, including healthcare, namely in the diagnosis of disease and ailments, and in the defense industry.

5.4 MEDICAL IMAGING (MRI, CT, ULTRASOUND, AND OTHER IMAGING TECHNIQUES)

Medical imaging in general has greatly benefitted from AI, especially from deep learning. Imaging techniques such as convolution neural networks (CNN) as well as other medical image analysis techniques find a list of applications in image analysis of patients at a high rate. It is a tradition as well as a practice in the medical profession that highly experienced physicians analyze medical images in order to detect, monitor, and characterize diseases. With the advent of AI, it has been now possible to quantify the complexities of the diseases evident in the images obtained through imaging techniques and later analyzed using AI [4].

AI benefits medical healthcare in terms of increased efficiency, as the number of experts in the field has reduced while the number of patients has increased and so has the associated data, thereby putting a strain on medical personnel as well as affecting expertise. AI compensates for this all by enhancing practice through automated intelligent algorithms.

AI inclusion reduces risk in medical treatment, automatically establishing routines to assist trained radiologists using the patient's images, which are scanned and identified for evidence of disease. New policies will help AI while including the latest technologies in medical imaging. AI assists in the process of assessment of the data in quantitative terms, which has been drawn from images which are later used for the detection, characterization, or monitoring of disease. Using computers, AI-assisted practices have transformed the field of radiology from a qualitatively based skill set to a quantitative domain which can be computed using machines.

5.5 CANCER DETECTION, DISEASE MANAGEMENT, AND DRUG DISCOVERY

The handcrafted engineered image features used in detection, such as tumor texture, have replaced mathematical equations through software. The input images are fed into machine-learning algorithms which are programmed to classify patients on the basis of image features and to base clinical decision making on these programs, which are not always the most optimal solutions as predefined features may not modify as per the variations in imaging characteristics in computed tomography (CT), positron emission tomography (PET) and magnetic resonance imaging (MRI), and there may be other associated issues such as signal-to-noise issues in the data.

Deep learning, on the other hand, has now gained an advantage in this process. Deep learning algorithms learn image features from data automatically, reducing the reliance on prior definition by human experts. Deep learning can quantify diagnoses and assist healthcare. Deep learning involves automated feature extraction and accurate segmentation of diseased tissues. AI-supported techniques such as deep learning are better at finding diseased areas in the body. However, there is an urgent need to acquire the right automation for replacing the traditional methods in segmentation of the diseased tissues.

Deep learning reduces the risk of undesired variation while covering a large variety of clinical conditions and parameters. It can identify image parameters and assess the importance of various elements which help to determine the best solution. Due to a growing number of application tasks in ultrasonography and MRI, improvements are on a par with radiologists' performance in detection and segmentation. AI-based deep learning has higher success when it comes to image sensitivities concerning the classification of lymph-node metastasis in PET-CT when compared with radiologists. The sensitivity vs. specificity tradeoffs are better supported by AI, as traditional manual methods have failed to satisfy clinical requirements, paving the way for an increased level of clinical utility.

5.5.1 AI in Oncology

Emerging AI technologies have the potential to have an effect on medical diagnosis via the assessment of the phenotypic characteristics in medical images, within the process starting with the detection of abnormality, followed by characterization and later on the monitoring of the impact. The tasks require both medical expertise for assessing and diagnosing patients and technical expertise to assist in image processing tasks.

In both areas mentioned above, emerging AI techniques may have great potential in the detection of cancers such as thoracic cancer using imaging techniques as well as methods such as mammography. State-of-the-art AI techniques can be easily integrated with existing clinical practices in detecting various kinds of tumors to assisting in treating these age-old diseases.

5.5.2 Detection Using AI

The detection of chronic diseases and tumors has traditionally been done using manual clinical trials, and these can now potentially be assisted by AI-based cognition, which confirms or rejects the finds reported by manual methods or vice versa.

Till now, the specialist in radiology specialists have assessed stacks of images, manually optimizing image parameters such as image planes, window width, and image level. Specialists employ their expertise, based on education strengthened through experience, to assess the patient's radiographic data.

They are trained to visually detect changes in image intensity or unusual patterns.

The subjective decision making of medical professionals on the array of these parameters is at the core of critical decisions that lead to a final opinion about the presence or absence of lung nodules, breast lesions, and colon polyps.

Assisted by computation, the automation of identification and processing techniques is called *computer-aided detection* (CADe). This was traditionally the radiologist's domain but now it is being transformed into a pattern-recognition domain assisted by computer vision algorithms which detect the image patterns.

These algorithms are limited in their scope to assess disease and imaging modalities and are effective on a case-by-case basis only.

As the accuracy of AI-based systems is not yet well established, the findings should always be assessed by radiologists to check if a specific annotation needs detailed investigation, which is a very labor-intensive and costly process.

While studying mammograms, radiologists are reluctant to change their decisions on the basis of the predefined, feature-based CADe systems, which show that AI has not been successful in impacting the performance of radiologists, which has been blamed for human error.

The advent of deep learning–based techniques in CADe has helped to focus on tumor detection in pulmonary nodules and prostate cancer, performed as multiparametric imaging techniques through MRI. Deep learning techniques have shown promising results, excelling as compared with other CADe systems and manual detection at both low sensitivity and high sensitivity.

5.5.3 Disease Characterization Using AI

Characterization is subdivided into the segmentation, diagnosis, and staging of a particular case of a disease, assessed in AI using quantification of the radiological characteristics of a disease through image size, image extent, and the internal texture of the area.

Radiologists are not capable of accurately determining various quantitative as well as qualitative features. The task is as complex as there are variations in these features on a case-by-case basis. AI has the ability to assess numerous quantitative as well as qualitative features along with their impact, which can be repeatedly true in nature. The complexity of this is due to the difficulty in differentiating between a malignant and a benign feature in the image when detecting lung tumors using CT scans, whereas it is easy for AI to determine these complex features, marked as imaging biomarkers.

5.5.4 Monitoring of Disease Using AI

The right decision, done well in a timely manner, helps in identifying critical solutions to chronic ailments. AI systems are part of the process called *workflow*. In what is known as the *pre-process* stage, it scans and registers the image multiple times; this is followed by the *diagnosis*, which involves the assessment of the image metrics. The data is compared using the metrics to quantify factors such as the tumor size in order to simplify it for experts as well as laypeople; this is often at the cost of misleading the public as to such things as isotropic tumor growth, which is difficult for humans to detect if it does not involve a large change in size, shape, or cavity. Any slight heterogeneity or texture change, poor image registration, the presence of multiple objects in the image, and physiological changes all make this task more challenging. Things get even more complex when inter-observer variability is involved, making it vulnerable and unreliable as it is an emerging field with much lower levels of adoption.

Multiple possible approaches, such as the automated registration of multiple images, are followed by subtraction of one image from another with the change being highlighted. The second method uses a pixel-by-pixel classification compared with a predefined set to create a complete mapping of the image data. The third method combines both the predefined and subsequent change sequence, which is prone to errors. Therefore the AI using deep learning architectures, such as recurrent neural networks (RNNs), can be used to assess the temporal sequence–based data formats and look poised to succeed in this area.

5.6 CHALLENGES FOR AI IN HEALTHCARE

There is a realization in the AI healthcare sector that these tools and techniques are finding a place within the field healthcare, where there is a willingness to integrate them with the computer-based toolset (especially when it comes to the deep learning toolset finding specific applications in clinical radiology). Moreover, a lively argument is floating atop questioning the real time needed to have AI find its place fully ranging from years onwards to decades or more. AI may find a natural role in healthcare applications where there is ample data available, especially in cases when the data is extremely difficult for health technicians and medical practitioners to decode or interpret, such as lung screening CTs, mammograms, and images from virtual colonoscopy. In addition, AI may also be a natural fit in areas where MRI and other imaging techniques are too complex to be understood by healthcare staff. This, however, will only be possible once the AI system has been further developed.

Data itself continues to be the critical element in learning AI systems as one out of four Americans undergo a CT test and one out of ten experiences an MRI examination each year, which increases the amount of image data generated per year. Digital healthcare systems in the United States, the European Union, and developing countries are focused on an effort to deal with the large amount of image data which need to be organized.

Huge amounts of medical data can be stored to facilitate ease of access and retrieval as data is always being lost, creating a major challenge in targeting an AI model.

5.7 FUTURE PROSPECTS

Advancements in the field of computer science, especially fields such as data science, machine learning, deep learning, and AI, have impacted the current state of healthcare. The most useful impact of neural nets–based deep learning can be seen in healthcare applications such as blood and urine data analysis and interpretation of medical images concerning vital organs. Due to such an impact, investment in these newly emerging technologies has become a necessity in the scientific community.

Imaging techniques have always been at the forefront of this effort, from X-rays and CTs to MRI and PET scanning. Medical imaging has evolved to focus on image quality, image sensitivity, and image resolution and can expose the smallest tissue abnormalities that cannot be detected by trained personnel or traditional AI techniques except for deep-learning algorithms, which are better poised with data promising improved performance, enhanced accuracy, and time savings.

There is an urgent need to develop the toolset which can give a realistic assessment of AI for patient welfare in terms of reproducibility and generalizability, a global toolset for quantifying patient data and setting widely agreed upon patient-related performance metrics as well as setting the standard for the imaging laws and bylaws as well as the standard formats to be used in experimentation.

In contrast with human intelligence, AI is better in one area at the cost of excellence in other areas, which issues a word of caution on all futuristic AI methods. AI has become too narrow in its focus on one task at a time where its performance barely passes human intelligence. Its excellence in data interpretation lacks the higher-level know-how in the way the human brain interprets information, and generalizations cannot be made for future cases. Therefore, AI cannot replace radiologists.

AI is not mature yet, so there is a real need that this over-stated excitement be replaced with realistic assessments and expectations, as radiologists will remain necessarily involved for a long time, and will be more involved as the systems become data-dependent and more technological, which will result in the most critical tool for training and developing logic-based effectiveness. Once AI surpasses humans, it will become an invaluable educational resource, while humans can only assist in interpreting the logic behind AI decisions in order to confirm their validity and detect any unseen data.

While AI algorithms come in dedicated software, deep-learning codes are open source and free, which encourages experimentation on a larger scale. There are also some signs of change as AI is set to shift from images to raw data, down-sampled and optimized to be understood by humans.

The simplified form and the lack of significant data can be dealt with if machines take over humans, as they are free from human errors. The more data there is, the more noise can be expected to be introduced, and the more stringent the algorithms must be. We must shift our AI algorithms to unsupervised learning techniques to work with stores of unlabeled datasets. There are pertinent questions about the ownership of and responsibility for AI policy and actions that follow, and this may weaken the current drive to implement this technology.

Other challenges include interoperability across the host of AI applications which spread across as the range of strong toolsets for inference and training, supporting identified patient data at a global level and helping the AI help patients across vast demographics, geographic regions, diseases, and healthcare standards. This will ensure the social responsibility of AI to benefit the multitudes within society.

5.8 HEALTHCARE IN 2019 AND BEYOND: FIVE AI TRENDS TO WATCH

AI has shocked the industry by the extent of its influence. The abilities of AI are wide-ranging, from simple to very complicated healthcare tasks, thereby enabling the former as an agent of change in the healthcare industry. AI has become firmly established as a revolutionizing element in the field. The overall revenue generated by the collaboration between healthcare and AI assisted by the global IT industry has been forecast to reach an enormous $223.16 billion by 2023.

Inspire of all this noise and excitement, it must be admitted that the inclusion of AI in the healthcare industry is largely vertical and sporadic and mainly in its infancy, which is different from other industries. There is great excitement in the industry, and the following are the major directions that the healthcare industry will take in 2019 and beyond.

1. Internet of Medical Things (IoMT)
2. Telemedicine
3. Wearables in healthcare
4. Chatbots
5. Cloud computing

5.8.1 THE INTERNET OF MEDICAL THINGS (IoMT)

The application of Internet of Things (IoT) in healthcare is called the Internet of Medical Things (IoMT). The total investment in the IoT healthcare market is expected to touch $136.8 billion across the world till 2021. Following are the areas that benefit from IoMT:

1. Drugs can be managed using IoT.
2. Diseases can be diagnosed and treated using IoT.
3. Remote monitoring of chronic diseases is improved through IoT.
4. IoT has helped to improve patients' experiences.

5.8.2 Telemedicine

This field has been around for many years under different names. With the adoption of digital technologies such as the vast usage of the mobile phone and the internet, the healthcare industry has been transformed over the last 20 years.

Healthcare is supported by these technologies in the evaluation, diagnosis, and treatment of patients using everyday technologies such as mobile phones and computers, which allow patients to access medical facilities over large distances. One of the fields that represent a major portion of this technological revolution is mental healthcare, as mentally ill patients can have instant access to a therapist.

5.8.3 Wearables in Healthcare

Strangely, it was once necessary to visit a clinic to take a blood test, check blood pressure, or check weight. AI has made it possible for all of this, including ECGs and blood tests, to now be done easily by patients themselves at home. Automation in the healthcare industry has also made it possible to send health alerts to patients.

5.8.4 Chatbots

The health industry is very familiar with chatbots. The healthcare industry is preparing itself to come to terms with this new reality and fully realize the transformation toward chatbots. Chatbots assist the healthcare industry by making it economical, assisting doctors in patient appointments and letting nurses know how and when to monitor critically ill patients. Chatbots can even be directly involved in helping the patients via consultations, although this is controversial. Ultimately, the chatbots will serve to help the healthcare industry in years to come.

5.8.5 Cloud Computing

Cloud computing is yet another highly promising arena from which healthcare can benefit. Patients usually have to wait a long time before they can access medical reports, but now all their medical data can be uploaded to the cloud so that patients have instant access.

Overall, 2019 and the beyond will see a great change in the healthcare industry as AI establishes itself in the sector. This will make a great difference to patients as they can access help around the clock. Let's be ready for this new wave of technological innovation.

5.9 AI IN HEALTHCARE: AN INNOVATION AND OPPORTUNITY IN DISGUISE

AI has a significant impact with great potential for healthcare [5], as it has had very promising results so far. AI has the most promising potential for healthcare in areas such as clinical research, drug development, and health insurance. This all helps in being the most economical and helpful to patients. The combined investment in the collaboration between AI and healthcare is an incredible amount; by the year 2021, it is estimated to reach 6.6 billion USD! Moreover, there is a prediction that savings of 150 billion USD per year can be expected from the investment in AI in healthcare till the year 2026.

The abovementioned advantages will be expected to occur step by step, starting with the introduction of automation in surgical clinics doing precision surgery followed by preventive interventions using predictable diagnostics aided by AI, thereby changing healthcare as we know it. It may take very many years to reach full maturity, however. Due to the nature of existing dilemmas in healthcare, the inter-dependent elements of access, affordability, and effectiveness will always involve negative tradeoffs for the industry.

The AI has the power to solve age-old dilemmas and empower the industry with the nontraditional methods to decrease costs, improve treatment, and strengthen access, all in one stroke.

The solution to costly structures in the existing healthcare is to replace time-consuming tasks with machines which will enable patients to take care of their needs where possible. This in turn promises to benefit people in the long run by making them healthier.

The effects of AI are so positive that it will ultimately meet 20% of demand that cannot be met by traditional clinical approaches. Healthcare facilitators ranging from hospitals to drug manufacturers have noticed this change. This has resulted in the rush towards investment in AI, machine learning, and predictive analysis as well as data analytics for the healthcare industry starting from 2018 onwards.

The effect of all these innovations promises to be felt in the managerial aspects of healthcare, which deal with operation and maintenance. An analysis of around 300 AI case studies has shown that the effect of AI in healthcare will be felt in customer care, consumption, higher quality services and goods, and highly personalized and data-driven services and equipment.

AI-related changes in healthcare are less noticeable at the customer level as most of these changes take place behind the scenes, according to the experts in the field, such as Dan Housman, CTO at ConvergeHealth by Deloitte.

According to Housman, radiologists will be more knowledgeable about radiology reports, and so the patient may be the one who is benefitted ultimately as the initial diagnosis will become highly accurate.

There are some areas in which investment in AI will transform the field in the coming years, according to Gurpreet Singh, an expert in the field:

1. Digitization
2. Engagement

3. Diagnostics
4. Digitization
   This helps in reducing the overall cost involved as digital tools like mobile technologies and wireless as well as the internet reduces the overall cost of healthcare.
5. Engagement
   AI technologies have benefited patients as well as general customers, in that patients and customers can easily communicate with healthcare providers and engage with systems and services.
6. Diagnostics
   AI-assisted products, services, and solutions help patients, who will be diagnosed or advised on health through AI algorithms.

5.10 THE STORY BEYOND 2019

As the proof of the concept about AI in healthcare becomes more firmly established, the advantages of the same are increasing and the capitalists have started investing more funds, resulting in new challenges for the adoption of the systems as well as for customers.

Experts in the field can trust AI tools as new algorithms are entering the market and experts like to see these algorithms to be first tested on the patients in clinical trials. Most experts will be skeptical until there is a vast database of these clinical trials on the new algorithms to verify their results.

There is another challenge in terms of patients' attitudes towards the adoption of these new algorithms, as one-quarter of patients are reluctant to use these newly AI-powered health services and products, according to one survey. The patients are reluctant as they have serious doubts about the efficacy and the reliability of these newly developed AI-powered solutions.

Another challenge stems from the scalability of the adoption of these newly discovered AI-powered solutions. AI-powered devices and solutions, while tested on a limited scale, may not be adoptable across the wide range of patients in large institutions and in society at large. There is another challenge, and that results from the fact that the solutions are too costly for patients, especially the ones located in rural areas.

Taking stock of the situation, especially the large extent of the public and society involved, there is a likelihood of a back-and-forth movement on the scale while these new AI techniques are initiated in the healthcare market.

On the whole, experts in the field are positive that adoption of AI will take place smoothly, at least in the far future of patient care. With the increase in the demand for healthcare as the numbers of patients increase, there will be a shortage of health professionals, thereby pushing towards the adoption of AI, Housman says.

With the stage set for AI in healthcare, five trends are predicted to dominate the field in the future. These trends are expected to influence the big technological firms, forcing these companies to invest with large swaths of money in global-scale research.

In 2018, AI progressed in many areas, platforms, tools, and applications of the healthcare industry globally. Therefore, AI has influenced many organizations

AI in Healthcare

influencing education, healthcare, and other industries, which are working on influencing the masses by providing solutions that depend upon the use of AI in healthcare. There is a range of other applications of AI: self-driving cars; agriculture, which is using robotics in crop sowing and pesticide spraying; and many other areas. Technological firms such as Google, Facebook, and Amazon have keenly invested billions of USD in AI and machine learning, and 2019 will see the further expansion of the scope of the AI as mentioned in the following section [6].

5.10.1 AI-Powered Chips

A new wave of the specialized AI-powered chips have been produced by manufacturers such as Intel, Nvidia, AMD, and ARM. They all plan to manufacture the chips to enhance routine AI algorithms for specific healthcare applications. AI applications employing image processing need specific chips which support large image-processing capabilities. It is not possible for CPU-based hardware to tackle facial recognition tasks and object recognition applications, as these require large mathematical operations in parallel. These performance enhancements will benefit these tasks with great solutions as their specialized designs will tackle applications such as natural language processing (NLP), speech recognition, and computer vision.

5.10.2 IoT and AI

IoT is a disruptive technology which has affected various fields, and AI is no exception as it has merged with AI in order to have an impact on our lives. IoT applications in industry integrate with AI to develop problem detection, early detection, better production, and many other applications in the very near future. The inclusion of AI as deep neural networks will benefit industrial processes where the prediction of certain processes is needed. Likewise, autonomous or self-driving cars will reveal large applications of AI and promise to resolve traffic issues and challenges in big cities globally.

5.10.3 Automated Machine Learning

Analytics is the other name for automatic machine learning as the impact of the latter is felt on the former in the future, resulting in huge revolution across the business world. The training models always benefit the algorithms as the complicated challenges in the fields will be solved without complex training.

5.10.4 The Rise of Facial Recognition

The field has become multifarious and scandalous as a great misuse of personal data occurred through Facebook. Despite all this, facial recognition will grow as the biggest field in the arena as the applications are hugely wide-ranging. The readability and accuracy of AI applications using facial recognition has benefitted a lot from the use of facial recognition in defense, at airports, and in various banking sectors where public access needs to be secure. Research into and the implementation of facial recognition will ultimately benefit defense and security agencies in tracking hackers, terrorists, criminals, and rogue elements in society; and businesses which need to provide more personalized data services to customers.

5.10.5 Increased Automation

There is great concern among the general public whether AI will replace people with machines, thereby making the former jobless. The inclusion of AI will certainly automate healthcare, and the global industry will benefit from this. Amazon has been impacted by this as robots have been used to make the work faster and more accurate. Enhanced automation and processes are preferable as there is a decreased risk of human error. Consequently, there must be more research done in AI. Increased usage of AI has benefitted society with automation, but this has also resulted in more problems and challenges which need more research by humans to solve. So, while there will be decrease in the low-skilled human workforce, there will also be an increase in highly skilled labor.

Lastly, the application of AI has continued to expand at an unusual rate, supported by the AI which continues to integrate with technological trends such as IoT and machine learning, and supported further by the boost in research that is owed to Google and other technological giants. AI is ready to move forward and expand its influence at a global level to have an impact on industries and businesses around the world, thereby benefitting humankind.

5.11 LIMITATIONS

The greatest impediment to the adoption of these futuristic AI healthcare systems is a lack of sufficient research funding, especially in the developing world. In the developed world, the pace of research on these systems is better, but thorough research effort is needed whereby all related issues concerning the research community and the findings must be clearly addressed, as traditional medical practices and healthcare infrastructure are based upon very old if not obsolete routines and practices which sometimes do not allow for the fast growth of AI in the manual or semi-automatic machinery and health infrastructure. In addition, an effort is needed on the part of researchers to conduct a thorough review of issues in the field through interviews with experts in the field in addition to consulting secondary sources of information.

The data gathered and analyzed through these interviews with experts could help with retrieving patient data from the digital machines as well as with the logical decision-making that these machines must do while assessing the patients' data. Therefore, the detailed understanding of the AI in healthcare is possible through study of the field data concerned as well as the knowledge of the logic flow and the algorithms employed while keeping in mind the data gathered on a human-to-human transfer basis.

The most important element is the robustness of the AI systems, it being user friendly, and that it has no bugs while processing the data, while the traditional infrastructural impediment reduces the chances of commercialization that AI may have already attained.

AI in Healthcare

On the other hand, keeping that all in mind, the current pace of AI in healthcare promises to transform it from an academic exercise into an important component in the health industry, with the immense promise of AI to become a major contributor in healthcare. In line with that, the most complex activities in the medical field are already being supported by DSS and are helping faculty and students to interpret the most complex medical riddles.

The current success rates reported by companies dealing with AI healthcare systems may be a major draw for both investors and customers, thereby giving a major boost to these developed technologies. This may also result in major investments by these companies and drive AI in the healthcare industry. The increasing trend toward the adoption of AI in healthcare may be visible as medical devices are in wide usage and it ensures the future success of AI in health. The areas of disease management and drug discovery have also benefitted from this larger effort. AI has the potential to benefit humankind but it needs to achieve the required level of robustness, and existing policymaking has to change a lot. The challenges are there as the existing toolset is not optimized enough and neither is the current state of AI. The industry, academia, and the government have to invest their own time and energy in improving tools and techniques to be on a par with the requirements of the medical community and patients [7].

5.12 CONCLUSIONS

There is an ongoing global effort to integrate existing healthcare with AI. The research effort has focused on the algorithms, which have initially centered on machine learning. Later, the algorithms moved to supervised learning followed by an effort to include the unsupervised learning toolset. Deep learning has evolved and shows great potential to aid AI at a global level. A range of startups have emerged in response to the global research effort which follow research trends. Data analytics based on machine and deep learning have remained at the forefront, with machines taking over routine jobs that require repetition and logic-based decision making. Deep learning has benefitted medical imaging, namely MRI, CT, ultrasound, and other imaging techniques in detecting tumors and other chronic diseases.

With this all said, there are obvious challenges associated with AI as it has become quite popular and reached its apex level. For example, self-driving cars need to slow down at a STOP sign, but instead of slowing down, the cars accelerate into busy intersections, raising huge concerns about AI flaws. An investigation has shown that the STOP sign has four small rectangles drawn over it, which confuse the car's AI, leading it to misread the STOP sign as a 45 speed limit sign.

Such an event must be avoided at all costs as it has the potential to sabotage AI systems. Researchers have already shown how AI can be fooled by misreading, misinterpreting, and by carefully misplacing the signs on the site. Facial recognition and other AI system algorithms can be misinterpreted by attaching a printed pattern on the attached glass or screen or on hats. The AI speech recognition system may also be fooled by playing around with its ability to make it sense phantom phrases by just by embedding white noises in the speech.

All these examples show how AI can be tricked and fooled and how a hacker or saboteur can break into AI systems for pattern recognition widely known as DNN.

REFERENCES

1. Bach Xuan Tran, Giang Thu Vu, Giang Hai Ha, Quan-Hoang Vuong, Manh-Tung Ho, Thu-Trang Vuong, Viet-Phuong La, Manh-Toan Ho, Kien-Cuong P. Nghiem, Huong Lan Thi Nguyen, Carl A. Latkin, Wilson W. S. Tam, Ngai-Man Cheung, Hong-Kong T. Nguyen, Cyrus S. H. Ho and Roger C. M. Ho. Global Evolution of Research in Artificial Intelligence in Health and Medicine: A Bibliometric Study. https://www.mdpi.com/2077-0383/8/3/360/pdf.
2. Emily Kuo. AI in Healthcare: Industry Landscape. https://techburst.io/ai-in-healthcare-industry-landscape-c433829b320c.
3. Artificial Intelligence, Applications in healthcare, Prasanna Vadhana Kannan, Research Analyst, Frost Sullivan, Singapore. https://www.asianhhm.com/technology-equipment/artificial-intelligence.
4. Ahmed Hosny, Chintan Parmar, John Quackenbush, Lawrence H. Schwartz and Hugo J. W. L. Aerts. 2018. Artificial intelligence in radiology. *Nat Rev Cancer* 18(8): 500–510. doi: 10.1038/s41568-018-0016-5 PMCID: PMC6268174
5. AI and Healthcare: A Giant Opportunity Intel AI Insights Team Insights Contributor FORBES INSIGHTS WithIntel AI| Paid Program Innovation. https://www.forbes.com/sites/insights-intelai/2019/02/11/ai-and-healthcare-a-giant-opportunity/#479bb89c4c68
6. Healthcare: 5 AI Trends to Watch for in 2019, It's Time for Healthcare to Catch Up, Ankita Mallick. https://dzone.com/articles/healthcare-5-ai-trends-to-watch-for-2019
7. 5 Top AI Trends COGNITIVE WORLD Taarini Kaur Dang Contributor COGNITIVE WORLDContributor Group Innovation 14-year old VC, Speaker, Author and Women Empowerment Champion. https://www.forbes.com/sites/cognitiveworld/2019/04/25/5-top-ai-trends/#4ccc5a476aa0

6 Security Vulnerabilities in the IoT

Neeraj Kumar Jain, Preeti Mittal, and Rajesh Kumar Saini

CONTENTS

6.1	Introduction to IoT	94
6.2	Background	95
6.3	Advantages of IoT	96
6.4	Disadvantages of IoT	96
6.5	Reasons for IoT Devices' Vulnerabilities	96
6.6	What Are IoT Devices and Services?	98
	6.6.1 IoT Devices	98
	6.6.2 Services of IoT	99
	6.6.3 IoT Devices and Services: Security	100
6.7	Security and Safety Threats, Attacks, and Vulnerabilities	101
	6.7.1 Vulnerability	102
	6.7.2 Exposure	103
	6.7.3 Threats	103
	6.7.4 Attacks	104
6.8	Primary Security and Privacy Goals	107
	6.8.1 Confidentiality	107
	6.8.2 Integrity	109
	6.8.3 Authentication and Authorization	109
	6.8.4 Availability	109
	6.8.5 Accountability	109
	6.8.6 Auditing	110
	6.8.7 Non-Repudiation	110
	6.8.8 Privacy Goals	110
6.9	Intruders, Motivations, and Capabilities	110
	6.9.1 Tools and Techniques Used by Intruders	111
	6.9.2 Purpose and Motivation of Attack	111
	6.9.3 Classification of Possible Intruders	112
	6.9.3.1 Individuals	112
	6.9.3.2 Organized Groups	112
	6.9.3.3 Intelligence Agency	113
6.10	Discussion and Conclusions	113
	6.10.1 Discussion	113
	6.10.2 Conclusions	113
References		113

6.1 INTRODUCTION TO IOT

The popularity and latest developments of the Internet of Things (IoT) [1] and its capabilities to serve in many ways have proved it the quickest upcoming technology, with a vast and comprehensive impact in improving lifestyle and business opportunities. IoT has slowly impacted almost all walks of life, from the biomedical and healthcare industries to training and business opportunities. There is a need in these sectors to store and process sensitive and confidential information about people and their ventures, financial transactions, development strategies, product marketing campaigns, and so on.

This huge interconnection of associated gadgets in the IoT has drawn attention to the need for strong security in light of the developing interest of millions and billions of associated devices and administrations around the world [2–4].

The extent of the risk is growing on a daily basis, with chances of attacks increasing in both number and unpredictability. Consequently, if IoT is to reach its fullest potential, foolproof security against vulnerabilities needs to be maintained [5].

Security is characterized as a procedure to protect objects against unapproved access, unauthorized use, physical harm, theft, and so on by maintaining a high level of privacy and trustworthiness for data related to an item and its accessibility at whatever point it is required [6].

Virtually, most connected objects, tangible or intangible, are vulnerable, as none of these objects can be both secure and useful. Whatever is used a lot offers crucial opportunities for intruders and hackers. A device in IoT can be kept safe if the process using it can fix its largest built-in value under certain conditions. Safety and security expectations in the IoT environment are same as in other ICT applications. So, to ensure the safety of the IoT infrastructure, a certain maximum built-in value of all the tangible (connected devices) and intangibles (services, information and data) needs to be maintained.

In this chapter, we will discuss and contribute a better knowledge and understanding of threats and cyber risks and their attributes encountered from many intruders, whether organizations or other intelligent bodies.

A system and infrastructure which is capable of identifying risks to the systems and their vulnerabilities (shortcomings in the system) are a must to ensure a comprehensive, robust, and complete set of safety measures and also helps to verify if the proposed safety mechanism is foolproof and secure from all possible threats and malicious attacks on the system (Figure 6.1).

To protect an IoT system, it is also mandatory at the same time that users, governments, and IoT engineers and developers must have clear vision and knowledge about the risks or threats and be able to answer the following questions [17]:

- What are the resources to be kept secure?
- Who are the major stakeholders?
- What are the actual and dangerous threats?
- Who is posing these threats?
- How strong are the threats that actors may impose in terms of capability and resource levels?

Security Vulnerabilities in the IoT 95

FIGURE 6.1 IoT smart devices under threat [7].

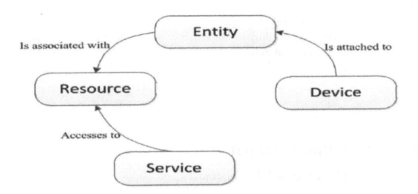

FIGURE 6.2 IoT model: key concepts and interactions [8].

- Which asset is more prone to which threats?
- Currently, are we safe? Is the current infrastructure protected from risks?
- What other safety measures can be implemented to better tackle threats?
- What are the threat avoidance and prevention methods?
- How do we recover from dangerous situations?

6.2 BACKGROUND

IoT is undoubtedly a very useful feature of the internet for communicating to real-world connected devices over distances [1]. Major IoT components, like devices, services, and networks, are important for successful and efficient operation. These are shown in Figure 6.2 [8].

In order to understand the correct meaning and functionality of the terms used above (IoT entities like devices, services, networks, etc.), it is useful to know the differences between similar terms. Figure 6.1 shows the relationship and association among various components of IoT in a better way [8] (Figure 6.2).

IoT entities may include a person, animals, automobiles, electronic appliances, and ready-to-sell goods or environments. Communication among these entities is established with the help of suitable hardware parts, known as *devices*: sensors, medical equipment, smartphones, radio frequency identification (RFID) tags, and so on. These devices play a vital role in connecting entities to the digital world.

In the present era of innovation, machine-to-machine (M2M) communication is the most common and popular application of IoT. It is currently being used in the power sector, smart transportation, the biomedical and healthcare industries, smart administration, water, oil, retail management, and various businesses to monitor and control clients, hardware, and so on.

6.3 ADVANTAGES OF IOT

Some of the advantages of IoT in brief are:

- Faster response
- Automation and control
- Timely information, leading to better decisions
- Monitoring and prompt action
- Time savings
- Economic benefits
- Great efficiency
- Improved quality of life

6.4 DISADVANTAGES OF IOT

Some of the disadvantages of IoT in brief are:

- Compatibility issues
- High complexity in implementation and maintenance
- Privacy/security
- Safety
- Staff downsizing and reduced employment

6.5 REASONS FOR IOT DEVICES' VULNERABILITIES

It is clear that IoT applications are serving mankind at their fullest and hence are very beneficial. Still, all this convenience comes with one major issue. IoT devices generally lack of security as their engineers neglect this important aspect so as to be able to launch their product as early as possible to make money. Providing safety and security to these devices followed by rigorous testing is essential but very costly and time consuming. So they may avoid going through it, which makes these devices more prone to attack and being exposed to hackers as an easy target. The following are some examples of problematic situations that may arise:

- A notorious hacker alters the medical prescription of a critical patient who is a very very important person (VVIP).
- Someone gets access to your mobile bank account.
- A departmental store automatically sends you an alternate product that you are allergic to, or food of a flavor which you dislike, or a product which is already expired.
- A hacker takes control of road traffic signals.
- A hacker gains malicious access to control of a city's electricity and water supply.

There may be many other threats imposing new challenges that need to be taken very seriously.

The enormous success of IoT has unfortunately attracted the attention of hackers and people with bad intentions. The connected devices or machines exchange lots of real-world data, such as a patient's medical history of a patient, doctor's prescription, financial information, and so on, that can be highly much lucrative for people intending to misuse it. Such data and information may be extremely precious for cyber-attackers for use for many malicious purposes, while it may be very dangerous, even life threatening, for others if it falls into the wrong hands [9].

Most of IoT devices are capable of performing on their own and need no human intervention at all. This situation may present a very attractive opportunity for an attacker to obtain valuable information from these devices. These devices transfer data using wireless networks where intruders and other attackers may use eavesdropping and many other theft techniques to access confidential information.

Many IoT devices have limited functionality and so cannot support complex security mechanisms. Lack of computing capabilities and low power are often the main reason for this.

In such situations, these cyber threats and attacks may take place against any IoT resource and service, resulting in various possible outcomes:

- There may be severe destruction or interruption of system operations.
- Harm can be caused to the population.
- Significant financial loss can ensue for owners and other stakeholders [10].
- On domestic automation systems, control of connected equipment like air conditioning and other connected physical security systems can be lost.
- Data stolen from home sensors for heating and lighting may reveal that no-one is in the house.
- Infrastructure-related data stolen from any public system may adversely affect basic amenities like power supply, electricity supply, internet connections, traffic control, and many others for the residents of that area [9].

In this way, security, safety, and privacy issues are on the increase and threaten both users and providers. The issue must be taken very seriously as a single connected device if hacked will cause severe loss in many terms.

In today's scenario, there is no doubt that IoT technology is making life easier in our homes, workplaces, or business sites, but at the same time it is inviting serious threats to safety and security as well. To overcome these issues, providers and users must walk hand in hand and should take great care in the manufacturing and use of these devices.

6.6 WHAT ARE IOT DEVICES AND SERVICES?

In IoT, physical devices like wireless sensors, actuators, computer devices, and so on work together through the internet via installed software. These devices are connected to networked objects and transfer data to objects or people on their own, without human intervention.

In the next section, important IoT concepts like IoT devices and IoT services and their relationships will be discussed in detail.

6.6.1 IoT Devices

A smart hardware device that connects an entity to the internet so that it can interact with any other connected device to share information is generally called an IoT *device* [8]. It is considered smart and it can be a home electronic appliance, automobile, biomedical equipment or healthcare device, building, workshop, factory, or almost anything connected to internet and sensors. Its primary objective is to collect and share information (about temperature, moisture, movement, humidity, presence, pressure, or pollution, etc.), and to control actuators (like switches, displays, panels, motor-driven shutters, or anything that can be handled by a device) and embedded computers as well [11].

IoT devices constitute a system wherein each connected device communicates with other connected devices, whether to automate household equipment, industry, and business-related issues, or to share important sensor data to stakeholders like competing businesses or doctors. These intelligent devices are divided into the following major categories:

- Consumer
- Enterprise
- Industrial

Connected devices such as smart TVs, toys, smart speakers, wearables, and other smart electronic home appliances come under the consumer category.

Smart electricity meters, commercial security systems and advanced technologies for smart cities like traffic monitoring and weather conditions and forecasting, air conditioning, car parking, smart thermostats, smart lighting and smart security, and enterprise and industrial uses all are examples of industrial and enterprise IoT devices [12] (Figure 6.3).

These IoT devices communicate with other IoT devices and systems through cellular (3G or LTE), wireless, or WLAN technologies [5]. This categorization of devices actually depends on many things, as shown below.

- *Size*: Small or normal
- *Mobility*: Fixed or in motion

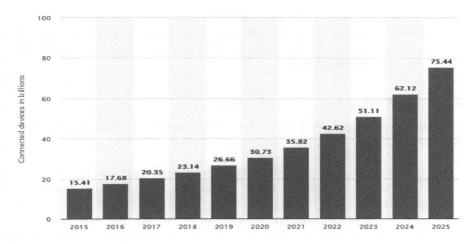

FIGURE 6.3 IoT connected devices worldwide from 2015 to 2025 (in billions) [13].

- *Location*: External or internal
- *Source of power*: Always connected or intermittently
- *Mode*: Automated or non-automated
- *Entity type*: Logical or physical
- *Connectivity*: Objects with IP or without IP

Selecting appropriate devices depends on the features of IoT devices, which are as follows:

- Capability to either actuate or sense or both
- Capability to ensure optimized use of power
- Capability to connect with the physical world
- Capability of discontinuous connectivity
- Mobility
- Security, safety, and privacy
- Rapid
- Trustworthy and reliable
- Available
- Needs protection/is self-serving

In actuality, IoT devices are often exposed to threats and need proper protection in order to retain their functionality. As these are connected to the Network of Networks (the internet), most of them are exposed to external and internal risks because of their characteristics. Making foolproof, safe, and secure IoT devices is a big challenge as there are many constraints like computational power, memory limits, and battery backup.

6.6.2 Services of IoT

These services are responsible for the integration and interconnection of IoT devices and other entities into the service-oriented architecture (SOA) world. IoT services is a collaboration among service providers and consumers. This helps in performing predefined functions to establish interaction with the physical world. It usually monitors and controls all the connected entities by taking initial actions to change the current status of entities [14].

6.6.3 IoT Devices and Services: Security

There are many reasons why IoT device security is more complex than IT security. Some of those reasons are as follows [15]:

- Establishing foolproof and sound security requires computing power, but due to their size constraints, IoT devices usually do not have such computing capabilities.
- Cloud data is more theft prone and easy for hackers to experiment with.
- MITM or man-in-the-middle attacks (in which an attacker secretly changes an actual conversation between two parties while they believe that everything is normal) are a well-known threat for IoT, and are hard to overcome.
- The complexity of IoT is also an invitation to possible attacks and creates much potential vulnerability.
- The cheaper prices of IoT devices and their easy availability on the market give an opportunity to hackers to study these devices thoroughly.
- Device information stored in the cloud encourages hackers to create bogus device identities.

The scope of IoT implementation is huge, but for the sake of safety and security, the following data security parameters will require maximum protection:

- Confidentiality
- Authenticity
- Availability
- Integrity

Securing and protecting both IoT devices and services from hackers means protecting devices mutually and externally. The ultimate objective is to save IoT Services, IoT devices (hardware resources), information and data at both stages, that is, at the time of transition and from storage.

Three main issues with almost all connected IoT devices and their services are as follows:

- Data confidentiality
- Privacy
- Trust

Let's discuss these one by one.

Security Vulnerabilities in the IoT

Data confidentiality is a major issue in IoT, in which both user and authorized devices can access data. Two important points to consider here are:

- Access control and authorization methods
- Authentication and identity management (IdM) systems

All IoT devices must verify and authorize the person or device before its use. This helps to ensure that after identification, only the authorized person or device can use the devices and services.

Access control means controlling access to devices and services by allowing or denying access based on access control rules, which should be clearly understood and relatively simple to create and manipulate.

In this way, authorization and access control are responsible for secure connection management between devices and services.

Authentication and identity management is critical in IoT, as every device(s), service(s), or multiple users need to be authenticated first through reliable methods. Identification of these players in a secure manner is also critical.

Privacy is another serious and important matter in IoT devices and services. Communicating entities are connected to each other and transferring data over the internet, which may be a threat to user privacy. Privacy in data management, like data collection, data sharing, and data security matters, needs special attention when creating security mechanisms.

Trust is an important dimension when establishing secure communication, especially when, in the IoT environment, multiple tasks are carried out over the internet. Trust needs to be developed among communicating entities and between users in the system and services.

Koien [6] says that trust in IoT devices is dependent on their components, as follows:

- *Power components*: Actual sources of power in use
- *Software components*: Operating systems, device drivers, hardware-based software, and various applications.
- *Hardware*: Processing units, storage, sensors, and actuators

6.7 SECURITY AND SAFETY THREATS, ATTACKS, AND VULNERABILITIES

We must understand our assets before going into the details of safety, risks, threats, attacks, challenges, and vulnerabilities. It is important to know everything that is a part of actual system like the inventory including IoT components and services. We actually need to protect the system from attacks.

The resources in an IoT system are valuable assets and sensitive in nature. Some important resources or assets are as follows:

- System hardware, infrastructure, and machinery
- Software components for individual devices

- Services offered by IoT systems
- Valuable data provided by the services

6.7.1 Vulnerability

A vulnerability is a weak feature of a facility which a threat actor can exploit for their own gain. Vulnerabilities means chances to get entry and to gain unauthorized access to assets and services with malicious intentions [16]. Vulnerabilities in assets means weaknesses in physical appearances, administration, procedures, monitoring and control, organization, personnel, management, hardware and software components, data, and information. These are prone to being accessed by hackers who misuse the system in order to affect business goals.

A vulnerability on its own is not harmful; rather it is just a situation that can invite a risk. It is important to reduce the vulnerabilities that result from multiple sources. Vulnerabilities will persist until the asset is changed so that the vulnerability no longer exists.

Improper access control mechanisms are an example of a vulnerability. Not all vulnerabilities invite attacks; only those which pose a possible threat require immediate action.

The solution to vulnerability issues is deep analysis, which is the close monitoring of aspects and features that are prone to attack by identified threats. A vulnerability in a system or resource is a risk that can be defined as the ease with which attackers can misuse a system or its asset. Normally it is defined in three classes as high (maximum), medium (moderate), and low (least), according to a vulnerability assessment.

In IoT systems, vulnerabilities may occur in different areas and may be the result of:

- Weakness in system hardware
- Weakness in software
- Weak and inefficient protocols and procedures
- Weaknesses in system users themselves [7]

Two major components of IoT systems are hardware and software, which are parts of the system that may have weaknesses in terms of design flaws.

Vulnerabilities due to hardware are complex to recognize and resolve, even in in cases where these are known to result from hardware compatibility and interoperability issues.

Vulnerabilities due to software usually originate in operating systems, utilities, application software, communication rules, and drivers for devices.

The major causes of shortcomings in software products are human error and software complexity. The following are the human-error reasons for unreliability and insecurity found in software products.

- Bad planning
- Incomplete software requirements
- Poor understanding of the product
- No conversation among developers and users

Security Vulnerabilities in the IoT

- Insufficient resources
- Lack of information and skills
- Inability to control the system

Due to these human weaknesses, software products are ultimately more prone to technical vulnerabilities [7].

6.7.2 Exposure

Exposure to the open world is a serious bottleneck in the system, as important and confidential information is always revealed and hence remains unsafe. This motivates and attracts attackers to carry out information-gathering activities.

In IoT, it is a serious concern that against exposure to physical attacks, people are resilient. Within many IoT applications, resources may be left less monitored and so are prone to be easily exposed and revealed to attackers. This kind of dangerous openness encourages attackers to take control of devices, access encrypted secrets, alter useful code, and, at worst, replace them with malicious devices [24].

6.7.3 Threats

A threat or risk can be defined as an action taken by attackers due to security weaknesses in the system and that hence is dangerous for system. Such possibilities of threats can appear in the system through human error or other sources like technical malfunctioning of devices or via natural threats [24].

Natural threats like fire, flood, hurricanes, drought, and earthquakes can destroy systems, but using backups or safeguards can safeguard the most important entity of the systems, that is, the data. Natural disasters are beyond human control, but these techniques are best used to ensure safe and secure systems.

Human threats like intrusion, hacking, and phishing are caused by humans with bad intentions. Such malicious threats may be internal or external.

Internal threats are generated from within the system; for example, someone has authorized access but misuses it.

External threats are generated from outside the system; for example, individuals or organizations working outside the network seek weaknesses in the system in order to harm it.

Human threats are classified into the following categories:

- *Unstructured threats* usually involve novice and mostly inexperienced individuals who try to breach system security using less harmful and easily available hacking techniques. These threats mostly involve unfocused assaults without any plan on one or more network systems. These attacks are made by humans with lesser integrity and excess time. These attacks may or may not be malicious, but they involve less chance of severe damage.
- *Structured threats* are more dangerous, being focused and planned by one or many individuals with advanced and high-level knowledge and skills and the latest hacking tools and techniques.

The targeted system is specially selected or detected randomly. The attackers in this case are experienced and possess knowledge of system topology, security, access points and procedures, and so on. They also create scripts or write malicious code to meet their objectives. The motivation behind such dangerous attacks is greed, international terrorism, politics, and government-sponsored attacks [18].

As IoT is becoming popular every day, a growing range of present devices have increased the number of cases of safety threats, with implications for the general public. Unfortunately, IoT comes with new challenges for security. There is a need for awareness about the new generation of smartphones, computers, and other devices that can be targeted with malware and are prone to attack.

6.7.4 Attacks

Attacks are malicious acts made by people usually known as hackers or intruders to take undue advantage of or to sabotage systems or to stop normal functioning of existing systems. They generally strive hard to exploit vulnerabilities using various tools and techniques. Attackers may be hackers, criminals, or other unauthorized entities (whether human or not) who are a threat to the cyber world. These attacks can be of different types. Let's discuss some of them in brief.

Active attacks: In active attacks, attackers or hackers modify information or generate bogus or fake messages. These attacks are in the form of interruption, modification, and fabrication.

Protecting systems from active attacks is a tedious task due to the wide range of vulnerabilities, like physical, network, and software vulnerabilities. Therefore, detection and recovery are better ways to handle these attacks [20] (Figure 6.4).

Passive attacks are attacks where the attacker uses unauthorized eavesdropping, simply monitoring and gathering information without altering it. An eavesdropper never alters data or the system.

Unlike an active attack, it is difficult to identify a passive attack as it doesn't involve any changes in the data or system. There are no noticeable symptoms of a passive attack as the attacked entity does not have any indication of the attack.

Encryption and decryption are the best ways to prevent such attacks. In these techniques, the data is encoded in unintelligible language at the sender's end, and it is later decoded into understandable form (Figure 6.5).

FIGURE 6.4 Active attack.

Security Vulnerabilities in the IoT

FIGURE 6.5 Passive attack.

So when the message is sent, it is encrypted and hence not of much use to hackers. This is why, in passive attacks, prevention is better than detection. Passive attacks look for less secure open ports. The attackers keep searching for vulnerabilities, and when they find one, they can get inside and gain access to the network and system [20].

Some other common attack types are as follows:

- *Physical attacks*: These attacks are focused on hardware components and devices attached to the system.
- *Node tampering*: Here an attacker usually alters the compromised node and steals confidential and sensitive data like encryption keys and so on.
- *RF interference on RFIDs*: Here an attacker uses denial of service (DoS) by interrupting radio frequency signals during RFID communication.
- *Node jamming in wireless sensor networks (WSNs)*: Here, using jammers, an attacker interrupts the wireless communication, resulting in a DoS attack (Figure 6.6).
- *Malicious node injection*: Here an attacker injects a single or multiple malicious nodes among actual nodes connected in the system and then sends wrong information to others. These attackers first put in a duplicate of node B followed by inserting another malicious node(s). Both these nodes then attack together and the victim node cannot function. Now, watchdog nodes may be used to signal that the victim node is itself acting maliciously. The monitoring verification (MOVE) technique is used to prevent such attacks. It continuously checks the monitoring node(s)' results and identifies malicious behavior on the part of any of the connected nodes. Malicious node injection is also known as *Man in the Middle Attack*.
- *Physical damage*: Here, attacker(s) physically attack any of the IoT components. It results in a DoS attack.
- *Sleep deprivation attack*: Here, an attacker tries to shut down running systems simply by making them use more power.
- *Social engineering*: Here attacker(s) interact with IoT systems and alter actual users of the system. The aim is to get confidential and sensitive information for misuse.

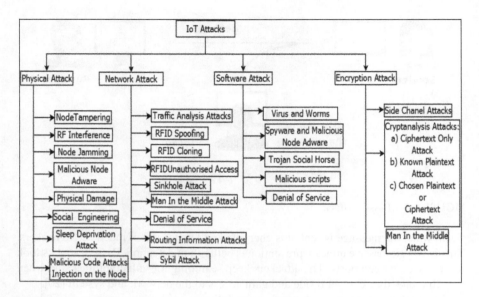

FIGURE 6.6 Possible IoT attacks [23].

- *Reconnaissance attacks*: In these attacks, an attacker or an intruder focuses on targeted system(s) to collect loopholes to fix the vulnerabilities. The aim is to get unauthorized access to systems, services, and vulnerabilities. For example, these attacks scan traffic analysis, network ports, and packet sniffers and continuously send fake messages to get IP addresses.
- *Denial-of-service (DoS)*: Here, attackers focus on controlling systems or network resources to make them unavailable to authorized people. Most IoT devices are likely to be vulnerable to this type of attack as devices have limited memory and computation capabilities.
- *Access attacks*: Attackers obtain access to networks or devices maliciously. In physical access attacks, intruders may get control over connected devices, while in remote access attacks, intruders operate on devices which are connected through IP addresses.
- *Attacks on privacy*: Due to the huge data and information exchange among connected devices over the internet, privacy protection is crucial. Some common and popular attacks on privacy are:
  - Data mining: Here attackers try to discover unexpected data and information that is difficult to find in specific databases.
  - Cyber espionage: Here attackers use cracking techniques, hacking tools, and malicious software to steal confidential data from organizations, individuals, or governments.
  - Eavesdropping: This involves listening to private communication between two parties or devices.
  - Tracking: Here attackers use the UID (unique identification number) of devices to capture user movements. Knowing users' locations helps them in many ways.

Security Vulnerabilities in the IoT 107

- Password-based attacks: Here attackers try to obtain a duplicate of a valid user password. This can be done as follows:
  1) Dictionary attack: Using letters and numbers in different possible combinations in order to obtain passwords
  2) Brute force attacks: Using cracking tools in order to check maximum combinations of possible passwords
- *Cyber-crimes*: Here attackers use smart devices and algorithms in order to exploit users and data for financial gains, like intellectual property theft, identity theft, and so on [24].
- *Destructive attacks*: Here attackers use space for destructive large-scale loss and destruction of life, money, and infrastructure; examples include terrorism and revenge attacks.
- *Supervisory Control and Data Acquisition (SCADA) Attacks*: SCADA systems are more vulnerable to cyber threats and attacks [24]. Two possible such attacks are:
  - Using denial-of-service (DoS) to interrupt functioning
  - Using Trojans and viruses to overwhelm (Table 6.1)

6.8 PRIMARY SECURITY AND PRIVACY GOALS

For successful and efficient IoT security management, the following things need to be properly addressed:

- Confidentiality
- Integrity
- Authorization
- Authentication
- Accountability
- Availability
- Auditing
- Non-repudiation
- Privacy goals

Let's discuss these important issues one by one.

6.8.1 CONFIDENTIALITY

Confidentiality is major security aspect in IoT which ensures that the transferred data is only accessible by the authorized destination, except in public domain systems, where an enormous amount of data flows freely. This may be unstructured data on social media as well. But in many scenarios, confidential data like a patient's medical history, a bank's financial data, people's credit card and bank account details, a country's army or military data, individuals' private conversations, server or email passwords, and so on must be kept safe and secure from unauthorized access.

TABLE 6.1
Comparison of Different Attacks [23]

Classifications /Parameters	Malicious Node Injection Attack	Sinkhole Attack	Worm Attack	Side-Channel Attack
OSI Layer	Physical	Network	Application	Application, Physical
Attack Type	Active —As the attacker compromises the node	Active —As it provides the wrong information, resulting in the packet dropping	Active —As it modifies the files	Passive —As the attacker can find encryption key by using the side channel information
Attacker Location	Both	External	Both	Internal
Attack Threat	Availability —Due to collision at the victim node it cannot transmit the packet	Availability, Confidentiality —As all the data is attracted to the compromised node	Availability, Integrity, Authenticity —As it can delete, modify the data	Confidentiality, Integrity —By using side channel information it can find the encryption key
Damage Level	High —As it can modify the data and pass the wrong info to other nodes	High —As all data is flowing through the compromised node, the attacker can do anything with the packet	High —As it can delete files, mail documents	High —As the attacker can obtain the secret key without being detected
Detection Chances	Low —As it is replica (clone) of legitimate node	Difficult to detect when it is near to the base station	Anti-virus can identify it	Negligible because the adversary uses side channel information
Possibility of Prevention	Yes —If we could avoid replication attack	Yes —If node authentication is provided	Yes —By avoiding suspicious sites, files	Yes —By using preventive methods
Attacks Based On…	Inserting malicious node	Routing	Malicious code	Side channel information
Vulnerability	Wireless nature and hidden node problem	Node authentication is not provided	Not following security policies	Side-channel information
Existing Solutions and Their Limitations	Not possible to detect if more than two nodes are malicious; consumes power because of overhearing [24]	When the malicious node is near to the base station (1 or 2 hop distance), the algorithm cannot accurately detect the sink hole node [24]	New worms are created every day	Affect the performance of the system

Security Vulnerabilities in the IoT

6.8.2 Integrity

Integrity is one of the most important security features to make sure that IoT services are safe and reliable for users. In IoT, the nature of systems working is heterogeneous with various integrity requirements [24]. An IoT system for remote patient caring and monitoring must have integrity to prevent random errors, as the sensitivity of this information is so high that any minor error in the communication may be life threatening for subjects [24].

6.8.3 Authentication and Authorization

Authentication and authorization are important and essential aspects in a security mechanism and are indispensable in the context of IoT. It is all about creating trust in connected systems, and there are many challenges in providing security for the IoT world. This is a way to provide efficient trust management for IoT systems, either locally centralized or globally distributed, by deploying authentication and authorization on the connected devices [21].

The increasing usability, popularity, and ubiquitous connectivity of IoT systems has generated this problem of authentication. Communication among IoT components takes place between:

- Machine to machine (M2M) or device to device
- Human to device(s)
- Human to human

The authorization process allows only authenticated entities to have access to services over the network.

6.8.4 Availability

In IoT, a connected device or user of a device must always be able to use the services at any time. To meet this challenge or to be available at all times, different hardware, software, and IoT devices must be capable enough to be available even when under cyber attack. Different systems have different requirements to ensure their availability; for example, emergency systems like fire, healthcare, or natural disaster monitoring need greater availability than moisture measurement or pollution sensors.

6.8.5 Accountability

Accountability is something that ensures that a system is safe as all other security techniques are functioning properly. It itself cannot prevent unauthorized activities or attacks but it can certainly help to ensure that the system is safe.

All security measures like integrity or confidentiality mean nothing if the system has no accountability. In the case of an attack, an identified and accountable entity must be available to check the actual cause and responsible person for it.

6.8.6 Auditing

A security audit is a proven method of measuring the safety and security of IoT device(s) or service(s) as it checks how well it adheres to the set protocols. As there are many loopholes and vulnerabilities present in systems, thorough audit for security becomes quite important to find all kinds of shortcomings and loopholes present in the system which may be exploited by hackers [24].

6.8.7 Non-Repudiation

Non-repudiation is the feature which ensures that entities, whether IoT users or devices, will always complete their actions and cannot deny functioning. Although this is not a very important security feature, it plays an important role in several specific systems:

- Payment systems: Either users or providers can deny payment.
- Email systems: A sender cannot deny sending a message, and a receiver cannot deny receiving it.

6.8.8 Privacy Goals

The privacy of an entity is the degree to which this will communicate and share information with its peers in both the internal environment and the external environment.

The key privacy objectives in IoT systems are:

- *In devices*: This depends on physical privacy. Confidential and secret information can be accessed only after theft or loss.
- *In communication*: This depends on the availability, integrity, and reliability of devices. Devices must be able to communicate as and when required while maintaining data privacy.
- *In storage*: To safeguard and protect data stored in devices, it is important to maintain privacy. A limited amount of data must be stored in devices and there should be a policy to erase all data from devices at the end of life. There should be a mechanism for the protection of user data in the devices. This data must be automatically erased if the device is lost or stolen.
- *In processing*: This is dependent on device and communication integrity. Devices must hide data from third parties.
- *In identity*: Only authorized entities must be able to see the identity of any device(s).
- *In location*: Only authorized entities must be able to see the geographical position of devices.

6.9 INTRUDERS, MOTIVATIONS, AND CAPABILITIES

A network intruder is someone who tries and succeeds to gain access to networked devices and services through physical, remote, and system attempts. The two categories of intruders are *internal* and *external* [22].

Security Vulnerabilities in the IoT 111

An internal intruder is a valid and authorized one having access to many areas of the internal system, services, and network, while an external intruder is one who is unauthorized for the system or network.

6.9.1 Tools and Techniques Used by Intruders

The following are tools and techniques commonly used by intruders to penetrate networks and breach network security [23]:

- *Trojan horse*: Trojans are legitimate software or malicious programs that are always trying to gain back-door entry to a computer system.
- *Virus*: A virus is a program with malicious code that implements undesired functions and is able to replicate itself to be inserted in other programs to do the same.
- *Worm*: A worm is also bad code that replicates and circulates itself to other computers on network with malicious intent. It uses the system network to spread from system to system, and hence it is more dangerous.
- *Vulnerability scanner*: Generally, intruders use this tool to find vulnerable systems in a network and identify weaknesses. Intruders also carry out port scanning to find open ports on the targeted computer.
- *Sniffer*: This tool finds passwords and other important data either within the computer or over the network.
- *Exploit*: taking advantage of known weaknesses in systems.
- *Social engineering*: Through this, confidential information can be obtained.
- *Root kit*: Intruders use this tool to hide the fact that a system has been hacked.

These intruders are a special kind of attackers motivated by financial gain, influencing public opinion, and spying. The role or motives of intruders differ from individual attackers to experienced and organized organizations [24].

6.9.2 Purpose and Motivation of Attack

Although it is difficult to predict the exact motives of the attackers in such intrusions, attack motives may be theft of identities or intellectual property, infrastructural access, and financial frauds [24].

Some other favorite targets for attacks include:

- Government websites
- Education websites
- Banking and financial systems
- News and social media websites
- Military networks
- Road traffic networks
- Public systems like infrastructure, electricity, and water supply systems
- Satellites and telecommunication systems

And there are many more.

6.9.3 CLASSIFICATION OF POSSIBLE INTRUDERS

The wo main types of intruders are internal and external [24].

Internal intruders are authorized people within the system with access to system, server or network while External intruders are generally not part of the system but try to harm the system from outside.

An individual attacker may attack on their own with small-scale objectives, while professional organizations may have bigger expectations. Let's look at the different kinds of intruders based on their capabilities, motives, and aims.

6.9.3.1 Individuals

Individual hackers work individually to target most vulnerable systems. Unlike professionals from spying agencies, they are less experienced and have fewer resources. As they usually attack relatively small systems with limited expertise and tools, the impact of the attack is less severe than ones launched by organized groups or spying agencies.

These attackers use social engineering techniques to steal basic information like addresses and passwords, and to attack the ports of the target system. Public and social media platforms are more prone to be accessed by hackers. Individual attackers also search for common vulnerabilities due to operating systems, PCs, and mobile phones.

Individual hackers also attack financial institutions like banks and credit-card information systems to alter the information in their favor. Due to the increasing trend of online shopping and payments, it has become easier for them to utilize stolen credit card details for shopping. These attackers prefer to use tools like viruses, sniffers, and worms to hack systems.

6.9.3.2 Organized Groups

Criminal groups and spying agencies are also impressed with the increasing popularity of ICT and IoT technology. They see various opportunities to gain advantage through hacking. The motivations of these groups may be surprising.

- Taking revenge
- Stealing business ideas from organizations
- Spying on economic systems
- Targeting the national information infrastructure
- and many more

These organizations tend to sell stolen data, such as the personal information of individuals or banking and financial information, to organizations, individual hackers, or terrorists, as well as to governments.

These organized groups are much capable in terms of resources. They can easily arrange money, expertise, and resources. They have the expertise to create malicious programs and denial-of-service (DoS) attacks.

Money is no object for these organized groups, so they can hire or buy the latest tools and everything else required. Considering it is a profitable business, such intruders can tolerate higher risks than individual hackers.

6.9.3.3 Intelligence Agency

Many countries have national intelligence agencies which continuously monitor the military and other important systems of different countries for different purposes like industrial, political, and military espionage (spying). Such agencies need to maintain a team of experts and research-oriented infrastructure to create these technologies in order to achieve their malicious intentions.

These types of intrusion expert agencies are the biggest threats to important systems in the national interest.

6.10 DISCUSSION AND CONCLUSIONS

6.10.1 Discussion

The internet is the backbone and soul of IoT, and thus all the security issues and threats that affect the internet will also affect IoT. Sensitive IoT nodes and devices with weaker capabilities and inadequate resources make safety and security issues in IoT very challenging. Additionally, the rapid development of and increasing demand for IoT devices necessitates the utmost security.

Due to processing limitations, capability and speed of these devices, traditional security measures are not sufficient for handling threats to IoT systems [23].

So IoT infrastructure must be built to ensure that using it is easy and safe, so that consumers can enjoy its advantages without worrying about security and privacy threats.

6.10.2 Conclusions

Like traditional network architecture for communication, IoT also uses network architecture, and thus is also affected by the shortcomings of traditional design and mechanisms. Along with the rapid expansion and adoption of IoT, various types of threats and attacks have been developed to breach the security barriers of IoT devices and services. Many scientists and researchers have suggested and proposed many ways to counter these attacks.

The implementation of security measures and techniques needs better computing power and device battery power, which most IoT technology and its devices lack. Therefore some lightweight, robust, and well-tested security mechanism is needed to ensure flawless and fearless use of IoT technology. Different types of attacks have been explained and classified here. Some of these threats and attacks may be handled by following security precautions, for example, implementing features like monitoring device identity and using safer devices which are difficult to tamper with at the time of communication. Still, for some known attacks which are not easily detectable, there is an urgent need to find and identify safe, secure, and better solutions [23].

REFERENCES

1. Atzori, L., Iera, A., and Morabito, G. (2010). The internet of things: A survey. *Computer Networks, 54*(15), 2787–2805.
2. Kumar, J. S., and Patel, D. R. (2014). A survey on internet of things: Security and privacy issues. *International Journal of Computer and Applications, 90*(11).

3. Stango, A., Prasad, N. R., and Kyriazanos, D. M. (2009, June). A threat analysis methodology for security evaluation and enhancement planning. In: *2009 Third International Conference on Emerging Security Information, Systems and Technologies* (pp. 262–267). IEEE.
4. Du, J., and Chao, S. (2010, August). A study of information security for M2M of IOT. In: *2010 3rd International Conference on Advanced Computer Theory and Engineering (ICACTE)* (Vol. 3, pp. V3–576). IEEE.
5. Taneja, M. (2013, October). An analytics framework to detect compromised IoT devices using mobility behavior. In: *2013 International Conference on ICT Convergence (ICTC)* (pp. 38–43). IEEE.
6. Koien, G. M., and Oleshchuk, V. A. (2013). *Aspects of Personal Privacy in Communications: Problems, Technology and Solutions.* River Publishers.
7. Abomhara, M. (2015). Cyber security and the internet of things: Vulnerabilities, threats, intruders and attacks. *Journal of Cyber Security and Mobility, 4*(1), 65–88.
8. De, S., Barnaghi, P., Bauer, M., and Meissner, S. (2011, September). Service modelling for the Internet of Things. In: *2011 Federated Conference on Computer Science and Information Systems (FedCSIS)* (pp. 949–955). IEEE.
9. Roman, R., Zhou, J., and Lopez, J. (2013). On the features and challenges of security and privacy in distributed internet of things. *Computer Networks, 57*(10), 2266–2279.
10. Rudner, M. (2013). Cyber-threats to critical national infrastructure: An intelligence challenge. *International Journal of Intelligence and Counterintelligence, 26*(3), 453–481.
11. Gluhak, A., Krco, S., Nati, M., Pfisterer, D., Mitton, N., and Razafindralambo, T. (2011). A survey on facilities for experimental internet of things research. *IEEE Communications Magazine, 49*(11), 58–67.
12. https://internetofthingsagenda.techtarget.com/definition/IoT-device.
13. https://www.statista.com/statistics/471264/iot-number-of-connected-devices-worldwide/.
14. Thoma, M., Meyer, S., Sperner, K., Meissner, and Braun, T. (2012). On iot-services: Survey, classification and enterprise integration. In: *Green Computing and Communications (GreenCom), 2012 IEEE International Conference On* (pp. 257–260). IEEE.
15. http://techgenix.com/iot-device-security-guide/.
16. https://www.sciencedirect.com/topics/computer-science/vulnerabilities.
17. http://www.iosrjournals.org/iosr-jce/papers/Vol19-issue6/Version-1/M1906018594.pdf.
18. https://medium.com/. @CybriantMSSP/types-of-network-security-threats-and-how-to-combat-them-b6624428b152.
19. Li, F., Lai, A., and Ddl, D. 2011. Evidence of advanced persistent threat: A case study of malware for political espionage. In: *Malicious and Unwanted Software (MALWARE), 2011 6th International Conference on* (pp. 102–109). IEEE.
20. https://techdifferences.com/difference-between-active-and-passive-attacks.html.
21. https://www.computer.org/csdl/magazine/it/2017/05/mit2017050027/13rRUwIF6ho.
22. https://core.ac.uk/download/pdf/25782486.pdf.
23. Deogirikar, J., and Vidhate, A. (2017, February). Security attacks in IoT: A survey. In: *2017 International Conference on I-SMAC (IoT in Social, Mobile, Analytics and Cloud)(I-SMAC)* (pp. 32–37). IEEE.
24. Abomhara, M. (2015). Cyber security and the internet of things: Vulnerabilities, threats, intruders and attacks. *Journal of Cyber Security and Mobility, 4*(1), 65–88.
25. https://pixabay.com/illustrations/communication-internet-1927706/.

7 Research on IoT Governance, Security, and Privacy Issues of Internet of Things

Manish Bhardwaj

CONTENTS

7.1 Introduction 115
    7.1.1 Birth of IoT 115
    7.1.2 Introduction to IoT 116
    7.1.3 Edges of IoT 116
    7.1.4 IoT Hardware 117
7.2 IoT across Numerous Domains 119
7.3 IoT Security Challenges 120
    7.3.1 Famous IoT Security Breaches and IoT Hacks 121
    7.3.2 IoT Security Apparatuses and Enactment 122
7.4 What Enterprises Are Most Helpless against IoT Security Dangers? 122
    7.4.1 Step-by-Step Directions to Secure IoT Frameworks and Gadgets ... 123
7.5 Common IoT Safety Efforts 123
7.6 Difficulties for Privacy and Security in IoT 126
    7.6.1 Security and Protection 127
7.7 Conclusion 127
Bibliography 128

7.1 INTRODUCTION

IoT is creating a huge system in which devices can be associated with each other and are able to interface with each other. This is driving mechanization to the next level, where devices can speak to each other and make decisions on their own, with no human intervention.

7.1.1 BIRTH OF IOT

This chapter is an educational exercise where we explore the expression "Internet of Things." The expression "Internet of Things" (IoT) was coined by Kevin Ashton in a presentation at Proctor and Gamble in 1999. He is a co-founder of MIT's Auto-ID research lab. He spearheaded radio frequency identification (RFID, utilized in standardized identification locators) for the assembly network which the executives house. He also began Zensi, a corporation that produces energy, detecting and checking innovation.

Kevin Ashton wrote the term in 2009 in his RFID report. This may assist you in comprehending IoT from its core. If we had PCs that knew it all—utilizing data they assembled with no help from us—we could track and tally everything, and thereby greatly reduce waste, misfortune, and cost. We would understand when things needed replacing, fixing, or reviewing, and whether they were new or past their best. We have to alter PCs' strategies for event data, so that they can see, hear, and smell the globe for themselves, within the totality of its discretionary brilliance. Ashton's statement provides inspiration as to the philosophies behind the advancement of IoT. In this chapter, we will strive to understand the developments of this term and try to comprehend IoT on a basic level. After this, we'll push ahead towards the benefits of IoT.

7.1.2 Introduction to IoT

The "Things" in IoT refers to any device with any form of inbuilt sensors with the ability to collect and move data over a system without manual intervention. The inserted innovation within the device encourages it to connect with interior states and therefore outer conditions, which therefore helps in selecting a method for the particular data type. More or less, IoT links devices online and allows them to speak with each other online. IoT could be a monster system of associated gadgets—which all accumulate and provide information concerning however they're utilized and therefore the conditions within which they work. Thus, all gadgets gain from the expertise of other gadgets, as do people. IoT is trying to increase the links between humans—i.e, humans get together, contribute, and work alongside one another. Perhaps this sounds somewhat lost, however we may better comprehend this with an analogy. A designer presents the developer with a proposal containing the measures, rationale, blunders, and special cases for a project. If there are any problems, a tester imparts them back to the developer. After several cycles, the project is complete. Likewise, an area temperature device accumulates data and sends it over the system; the data is then utilized by numerous device sensors, which switch their temperatures as desired. For an in-practice example, an icebox device will accumulate knowledge of surface temperature and will modify the cooler's temperature accordingly. Air conditioning systems can likewise modify temperature in the same way. This is often the way gadgets work together. This gives an idea of what IoT actually is. As we progress in our IoT educational exercise, we'll examine the benefits of IoT and therefore the instrumentation utilized in IoT applications with Figure 7.1.

7.1.3 Edges of IoT

Since IoT enables devices to be controlled remotely over the web, its data and probabilities solidly relate the physical world to the PC-based environment, which consists for the most part of structures using sensors and links. The interconnection of different

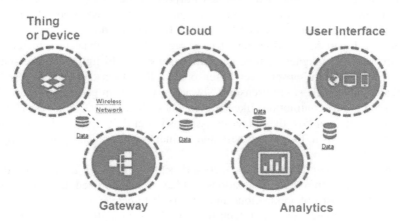

FIGURE 7.1 Major components of IoT.

devices enables computerization in wider fields and applications. Frequently, this conveyance improves precision, capability, and cost-effectiveness, with reduced human intercession. It consolidates developments, for example great structures, strong homes, smart transportation, and quick urban networks. The $64,000 benefits of IoT are:

- Improved customer engagement—IoT improves buyer skill by motorizing the movement of vehicles. For example, any issue inside a vehicle is perceived by the sensors. The driver, as well as the manufacturerer, will be alerted, and when the driver touches base at a repairs workshop, the manufacurer can ensure that new parts to replace defective ones are available at the workshop.
- Specific enhancement—IoT has sparked an incredible number of rising innovations. The manufacturer will accumulate information from various vehicle sensors and work on them to improve their set up and upgrade them.
- Diminished waste—On a rolling basis, IoT provides information that may be essential for the smooth running of businesses. For instance, if a producer finds a significant number of products are of inadequate quality, he will go to the factory where the product is made and address the trouble directly.

Nowadays, we use many IoT-approved contraptions that constantly gather information and transport it through different devices. As we continue forward, we ought to watch out for IoT applications. In like manner, take a guess at the IoT contraptions that we use most frequently in our everyday lives.

7.1.4 IoT Hardware

By and by, you begin to recognize IoT devices everywhere, and apply them in your lives. First you notice sensors which see the planet; then you need a dashboard to screen your data and show it in a clear and structured manner. Finally, you require a gadget with the ability you seek. The key purpose of the structure would be to recognize express conditions and act on them. One important task is corroborating the correspondence between the contraptions and the dashboard.

Some of the fundamental sensors that are regularly used are accelerometers, temperature sensors, magnetometers, separation sensors, whirligigs, picture sensors, acoustic sensors, light sensors, weight sensors, gas RFID sensors, dampness sensors, and scaled downscale stream sensors. Nowadays we will in general moreover have different wearable contraptions like smartwatches, shoes, and 3D glasses. These are the good example of how these devices assist us in our daily lives: 3D glasses adjust your perception of a TV's brightness and a smartwatch screens your step-by-step activities and welfare.

I feel the most important devices, colossally significant to IoT, are personal digital assistants (PDAs), moveable applications that have revolutionized the world of innovation. PDAs consist of applications and sensors that uncover a wealth of knowledge about their consumers: geo-area data; light conditions; the direction your device is travelling; and much much more. It in addition accompanies accessibility alternatives like Wi-Fi, Bluetooth, and cell data, that encourage them to talk with totally different gadgets. Hence, as a result of these default characteristics of mobile phones, PDAs are at the core of the IoT body. Today, smartphones collaborate straightforwardly with smartwatches and welfare bands and upgrade the consumer expertise and this is shown with the lifetime of IoT devices (Figure 7.2).

IoT utilizes totally different advancements and conventions to talk with gadgets. The many advancements and conventions include Bluetooth, remote, NFC, RFID, radio conventions, and WiFi.

IoT applications are successful in all businesses and markets. The IoT incorporates a vast range of applications over totally different enterprises. It ranges over gatherings of purchasers, as people worldwide try to decrease and ration energy in their homes to huge associations worldwide compelled to improve the efficiency of business. IoT has not simply made itself useful in basic applications, but has nevertheless

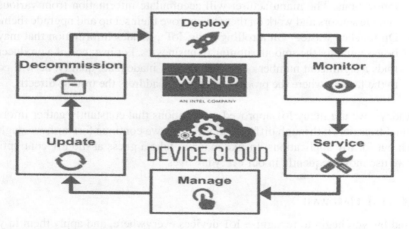

FIGURE 7.2 IoT hardware lifecycle.

Governance, Security, and Privacy Issues   119

additionally supported the concept of innovative mechanization that has risen every decade. We should understand that the skills of IoT cross over numerous enterprises and are reforming everything.

7.2 IOT ACROSS NUMEROUS DOMAINS

- *Energy applications*: People and associations are trying various approaches to innovate energy consumption. IoT provides an approach to not simply screen the energy used at the machine level, but also at the house level, lattice level, or maybe at the appropriation level. Good meters and energy grids are utilized to measure and transport energy. In addition, they identify dangers to the framework execution and security, thereby defending machines from harm.
- *Social insurance applications*: Smartwatches and welfare gadgets have changes the way people interact with their own health. People will check their data regularly throughout the day. Not only this, but if a patient is going to hospital in an ambulance, once the patient arrives at the emergency clinic, their health report is analyzed by specialists and therefore the clinic can quickly begin treatment. The data accumulated from numerous welfare applications speedily identifies health issues and hastens their fix.
- *Education*: IoT provides coaching that fills the holes within education. It improves the quality of coaching, yet enhances the expense, and improves administration by considering students' answers and execution.
- *Government*: Governments try to built cities utilizing IoT arrangements. IoT upgrades equipped power frameworks and administrations. It provides higher security between economical and elite gadgets. IoT encourages government offices to gather data ceaselessly and improve their services, for example citizen welfare, transportation, education, and so on.
- *Air and water pollution*: Through various different sensors, we determine pollution levels in air and water by taking continuous measurements. This helps to locate sources of pollution and anticipate connected debacles. IoT lets us limit human intercession in the investigation, especially when conditions are dangerous for humans. Frameworks distinguish changes in harvests, soil, pollution, and that is just the tip of the iceberg.
- *Transportation*: IoT has changed travel and transportation. Presently, we are creating auto-drive cars with movement sensors, traffic lights that notice the traffic and switch as a result, smart car parks, and so on. Likewise, various sensors in your car show you the statistics of your vehicle, so that you do not confront any problems when driving.
- *Promoting your item*: Businesses dissect and react to consumer inclinations by collecting data about the consumer experience and product purchasing preferences. This helps to constantly improve business procedures and performance. Ground-breaking IoT arrangements are astoundingly touching totally different areas, and we take a profound jump to understand Raspberry Pi, which is typically used to set up IoT arrangements. After this, we'll create an IoT application. IoT security was the topic of examination,

where a typical IoT appliance was used to invade and assault the larger system. Actualizing safety efforts is paramount to guaranteeing the safety of IoT systems and the gadgets connected to them.

7.3 IOT SECURITY CHALLENGES

Various difficulties forestall the certainty of security in an IoT domain. Since systems administration apparatuses are mostly new, security is not usually viewed as a prime concern in their development. What is more, in light of the fact that IoT is in its infancy, planners and manufacturers are increasingly keen on showcasing their products as quickly as possible, without troubling too much about security.

A noteworthy issue with IoT security is the utilization of hardcoded or default passwords. These might prompt security breaches no matter whether or not passwords are reworked, and indeed they're often not able to avert penetration.

Another common issue with IoT gadgets is that they're often quality-compelled and do not contain the assets vital to execute solid security. In this aspect, many gadgets do not or cannot detect breaches of security. For example, sensors that measure damp or temperature cannot trot out cutting-edge coding or safety efforts. Also, there is a "make it and forget it" outlook on IoT gadgets and their constantly evolving product lines; they are released, a finite number manufactured, and then left till they are discounted—they barely ever get security updates or fixes. From a producer's perspective, building security in from the beginning may be expensive, hinder company advancement, and cause the appliance to not work as it ought to.

Genealogy databases not naturally intended for the IoT network are another security challenge. Replacing heritage foundations with associated innovations is cost-restrictive, and a big range of benefits are retrofitted with shrewd sensors. Nevertheless, as inheritance resources haven't been overhauled for cost-effectiveness, or ever had protection of personal data, the scope for security breaches is huge.

Instead of permanent updates, various frameworks simply incorporate information for a time allotment, visible only to a specified group. For heritage and new resources, security will fail if extra information is not enclosed. Also, a wide range of IoT information stays within the system for an extended time beyond that agreed by the parties.

IoT security is lacks any industry-acknowledged benchmarks. While various IoT security structures exist, there's no single settled upon system. Monumental organizations and businesses might have their own specific tips, whereas, as an example, mechanical IoT engineers have entirely different norms from business pioneers, perhaps even contradictory ones. The wide and varied assortment of those norms makes it onerous to verify frameworks, not to mention guarantee the security between them.

This, along with operational innovation (OT) systems, has created many difficulties for security teams, notably those entrusted with guaranteeing frameworks and guaranteeing complete security in regions outside their domain of ability. There is the expectation that data is saved or deleted in accordance with users' agreements and rights, and IT teams with the most basic skills become answerable for IoT security.

Associations should decipher a way to see security as a mutual issue, from maker to specialist to end-user. Manufacturers and specialist organizations have to be

Governance, Security, and Privacy Issues 121

compelled to organize the safety and protection of their things, and what is more offer processes, testing, and approval. But the worry doesn't finish there; end-users should certify to avoid potential risk, together with evolving passwords, introducing patches once accessible, and utilizing security programming.

7.3.1 Famous IoT Security Breaches and IoT Hacks

Security specialists for a long time have cautioned of the potential danger of giant quantities of unbound gadgets related to the net, ever since the IoT plan originally started in the late 90s. Varied assaults have stood out as actually interesting, from fridges and TVs being used to send spam, to programmers hacking tablets and conversing with children. It's imperative to note that many of the IoT hackers do not specialise in the gadgets themselves, rather they use IoT gadgets as a passage into the larger system.

In 2010, for example, analysts uncovered that the Stuxnet infection was used to physically damage Iranian rotators, with assaults starting in 2006 but the essential assault happening in 2009. This is viewed as, in all probability, the earliest case of IoT assault. Stuxnet targets supervisory control and data acquisition (SCADA) frameworks in an industrial control system (ICS), utilizing malware to contaminate directions sent by programmable logic controllers (PLCs).

Assaults on mechanical systems are constant, with malware including CrashOverride/Industroyer, Triton, and VPNFilter, which specialize in defenseless OT and trendy IoT frameworks. In December 2013, a specialist at business security firm Proofpoint INC. found the primary IoT botnet. As indicated by the specialist, over a quarter of the botnet was comprised of gadgets apart from PCs, together with TVs, tablets, and gaming consoles.

In 2015, security analysts Charlie Miller and Chris Valasek loaded a foreign program into a landrover, changing the radio station, turning its windscreen wipers and air conditioning on, and preventing the accelerator from operating. They previously said they might also execute the motor, connect with the brakes, and incapacitate the brakes. Miller and Valasek invaded the vehicle's system through Chrysler's in-vehicle network framework, Uconnect.

Mirai, one of the most important IoT botnets thus far, initially maltreated author Brian Krebs' website and online virtual host (OVH) in September 2016; the assaults checked in at 630 gigabits per second (Gbps) and 1.1 terabits per second (Tbps), respectively. The following month, the domain name system (DNS) specialist with Dyn's system was targeted, which, along with Amazon, Netflix, Twitter, and also the *New York Times*, was inaccessible for a substantial length of time. The assaults invaded the system through purchaser IoT gadgets, together with IP cameras and switches.

Various Mirai variations have since occurred, together with Hajime, Hide 'N Request, Masuta, PureMasuta, Wicked botnet, and Okiru, among others.

In a 2017 notice, the Food and Drug Administration (FDA) cautioned the deep-rooted frameworks in radio implantable heart gadgets, which together with pacemakers, defibrillators, and resynchronization gadgets may be defenseless against security interruptions and assaults.

7.3.2 IoT Security Apparatuses and Enactment

Numerous IoT security structures exist, but there's no single industry-acknowledged standard thus far. If there were some kind of standard structure, it could assist organizations in developing IoT gadgets. Such structures are proposed by the GSM Association, the IoT Security Foundation, the economic web pool, and so on.

In September 2015, the Federal Bureau of Investigation discharged an open administration declaration, Federal Bureau of Investigation Alert range I-091015-PSA, which warned of the potential vulnerabilities of IoT gadgets and offered client insurance and barrier proposals. In August 2017, Congress announced the IoT Cybersecurity Improvement Act, which required any IoT appliance offered to the U.S. government to not utilize default passwords, not have well-known vulnerabilities, and supply an instrument to repair the gadgets. While it only technically applied to manufacturers creating gadgets for the government, it set a pattern for safety efforts among all manufacturers.

Also in August 2017, the Developing Innovation and Growing the Internet of Things (DIGIT) Act passed the Senate, but it is still, at the time of writing, anticipating House endorsement. This bill would need the Department of Commerce to assemble an operating committe and create a report on IoT, taking into account security and protection.

While not IoT-explicit, the General Data Protection Regulation (GDPR), discharged in May 2018, lays out data security laws across the European Union. These securities encompass all IoT appliances and their systems and IoT gadget producers must adhere to them.

In 2018, Congress gave the State of Modern Application, Research, and Trends of IoT Act, or the SMART IoT Act, to prompt the Department of Commerce to steer the IoT business and provide suggestions as to the safe development of IoT gadgets.

In September 2018, California state legislation body signed legislation called SB-327, a law that specifies security requirements for IoT gadgets sold across the state.

7.4 WHAT ENTERPRISES ARE MOST HELPLESS AGAINST IOT SECURITY DANGERS?

IoT security hacks will occur in any business, from smart homes to factories to vehicles. The seriousness of the assault depends on the individual framework, the data gathered, and probably the information it contains.

An assault weakening the brakes of a vehicle, for example, or a hospital records system, as an example, hacked to gain a patient's prescription medicine, may be unsafe. In like manner, assault on a refrigeration framework lodging prescription that's discovered by the wider IoT framework will demolish the effectiveness of a drug if temperatures vary.

Different assaults, in any case, cannot be thought little of. For example, an assault against smart door locks may probably allow a criminal to enter a home. Or, on the other hand, in numerous cases, as an example, the 2013 Target hack or different security ruptures, an aggressor may run malware through the wider IoT system—a

Governance, Security, and Privacy Issues

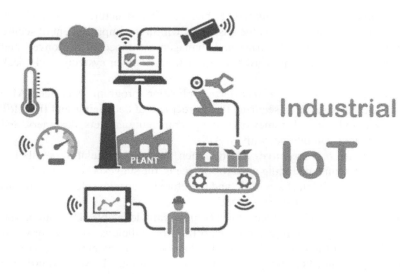

FIGURE 7.3 Industrial IoT.

HVAC framework, in Target's scenario—to delete data, unleashing devastation on those influenced.

7.4.1 Step-by-Step Directions to Secure IoT Frameworks and Gadgets

IoT security techniques are contingent upon your specific IoT application and your position within the IoT body. For example, IoT manufacturers—from item producers to semiconductor organizations—ought to specialise in structure security from the beginning, creating safe infrastructure, guaranteeing secure overhauls, giving bug fixes, and dynamically testing their products. A solution engineer must be employed to take care of secure programming advancement and secure identification. For those IoT frameworks, infrastructure security and validation are basic measures. Moreover, for directors, keeping up with the news, assuaging malware, examining, guaranteeing framework, and defensive qualifications are crucial as IoT industrialization is an important point, as shown in Figure 7.3.

7.5 COMMON IOT SAFETY EFFORTS

- *Consolidating security at the development stage.* IoT engineers must incorporate security at the start of any enterprise. Giving security the attention it deserves after all is basic, just as basic as using the most recent operating frameworks and utilizing secure infrastructure.
- *Hardcoded passwords must be used.* Operators need passwords in order for the device to be functional. On the off chance that a device has default passwords, buyers ought to refresh them utilizing a solid secret key or many-sided confirmation, or even bioscience, wherever conceivable.

- *PKI and advanced testaments.* Public key infrastructure (PKI) and 509 advanced declarations assume basic jobs within the improvement of secure IoT gadgets, giving the trust and management expected to convey and acknowledge open passwords, secure data trades over systems, and check temperament.
- *Programming interface security.* Application program interface (API) security is essential to securing the correctness of data being sent from IoT gadgets to back-end frameworks and guaranteeing gadgets, designers, and applications can interact with all APIs.
- *Personality the executives.* Giving each device a stimulating symbol is essential to understanding what the device is, the suggests how it works, the other devices it interfaces with, and therefore the correct safety precautions that must be taken for that gadget.
- *Equipment security.* Gadgets must be physically locatable. This can be significantly important if gadgets are utilized for malicious purposes or wherever they will not be used physically. Passwords are essential to verifying correspondence between gadgets. Data in transit must be verified utilizing algorithms.
- *System security.* Securing an IoT system means guaranteeing port security, handicapping ports causing issues and handicapping ports once not required; utilizing anti-malware, firewalls, and interruption discovery framework/interruption aversion framework; obstruction unapproved scientific discipline addresses; and guaranteeing frameworks are mounted and forward-thinking.
- *System access management.* The North Atlantic Council will facilitate, distinguish, and stock IoT gadgets associating with a system. This may provide a pattern for following and checking gadgets. IoT gadgets that require connection with the net must be divided into their own systems. System sections must screen for odd movement, and where the move occurs, should one be noticed.
- *Security passages.* As a passage between IoT gadgets and therefore the system, security doors have inbuilt power, memory, and capacities apart from the IoT gadgets themselves, which provides them the capability to actualize highlights; for instance, firewalls ensure hackers cannot get to the IoT gadgets they connect with. Updating gadgets and programming either over system associations or through cybernation is essential. Testing for vulnerabilities is in addition imperative to updating gadgets as quickly as time permits. IoT and operational framework security can't use old security networks. It's essential to remain up-to-date with the most recent new or obscure frameworks, adapt new styles and programming dialects, and be ready for brand new security challenges. C-level and cybersecurity teams must be ready for and responsive to current dangers and safety efforts.
- *Coordinating teams.* As well as being prepared, coordinating varied and experienced teams is useful. For example, having programing engineers work with security experts will facilitate simple controls and good security on gadgets throughout the advancement stage.

- *Customer instruction.* Consumers should be made aware of the threats of IoT frameworks and given instructions to stay secure; for instance, renewing default passwords and applying programming updates. Consumers will likewise demand device producers to create secure gadgets, and decline to utilize those that do not satisfy high security requirements.

Administration, security, and protection are inherently interlinked and they ought not be examined in a detached manner. We have a tendency to connect these aspects to ethics. Action Chain 05 doesn't characterize government arrangements, rather it centers upon analysis (which might in the long-term be utilized to assist methods or institutionalization exercises). Administration, security, and protection are the foremost testing problems within the Internet of Things and they are widely discussed in the literature. In the following sections, we'll condense the main functions of these three elements of the IoT as indicated by the first commitments projected in writing. The concepts of IoT governance, security, and privacy are likewise not utterly separate and are put forward by numerous government trade and analysis associations.

Inside the EU, "administration" alludes to the rules, procedures, and conduct that influence the style by which matters are organized, particularly as regards openness, support, responsibility, viability, and soundness. These five "standards of fine administration" fortify those of lower status and proportion. Governance is connected to the internet and there are as of now associations like IETF, ICANN, RIRs, ISOC, IEEE, IGF, and W3C, that are very capable and manage a different zone.

While these associations chip away at at internet administration, a wise advance is to expand these concepts to IoT administration. The difficulty is that the wide variety and heterogeneousness of advances and gadgets within the IoT need considerably more and more specific governance arrangements and methodologies.

Size and heterogeneousness are the two principle elements that influence the administration of IoT: administration is considered a double-edged blade, since it offers steadiness and backing for decisions, yet it likewise may lead to an over-controlled situation. This underlines the difficulty of finding a means of IoT administration besides the varied places of various partners: it's by all accounts untimely to start a discussion and there's no benefit to discovering distinctive standards for IoT administration problems that are isolated from common regulations. All things considered, since there are not any lawful structures for IoT administration, in spite of whether or not the differences between the IoT and the web were overestimated from the start, an investigation of the foremost IoT administration problems (authenticity, straightforwardness, responsibility, ill-conduct) would be advantageous. Except for strategy or ethical issues that impact administration itself, the discussions in this section explore specialized arrangements that currently exist.

Heterogeneity expects security to defeat the problem of actualizing effective conventions and calculations on each gadget over the various areas of IoT applications. While not experienced in security, managers are most likely not waiting to embrace IoT arrangements on a large scale. Therefore, the advance of regulations assists ability to anonymize clients' data and to allow data insurance are key parts.

In the IoT setting, it's arduous to isolate the concepts of governance, security, and privacy, on the grounds that tending to protection and security to accomplish trust in IoT would doubtless require administration systems too. As observed antecedently, due to the rising frequency of collaboration of IoT with consumers, ethical viewpoints cannot be ignored in administration, security, and protection. Here, we have a tendency to embrace the meaning of security displayed wherever security, data insurance, and data protection are reciprocal conditions for IoT administrations. Specifically, data protection aims to avoid wasting the confidentiality, integrity, and availability (CIA) of knowledge.

In Europe, some underlying work has been performed in relation to Regulation 611/2013, Article 4(3) in regard of creating a representative summary of appropriate innovative insurance measures. One noteworthy wellspring of this essential work were the reports on bioscience measures to verify personal data discharged by ENISA.

The rise of remote gadgets compounds the problem of security thanks to applications which gather data like legal names and contact details that may jeopardise consumers' privacy. As a result, protection must be upgraded so to limit the gathering of such data, hide important data from unauthorized parties, and send the correct data to approved groups. The administration of various gadgets, applications, and conventions is likewise attended utilizing the standards of administration, accomplishing a remarkable ability in numerous degrees of the IoT engineering.

Another important issue is that IoT is used in an exceedingly controlled manner: the creators provide instructions about each aspect, which seems, once all is said and done, the greater portion of the multifaceted nature bestowed by the wide methodology within the arrangement of administration, security, and protection policies. As it is, they also demonstrate a number of blessings accomplished utilizing the circulated methodology, by way of ability and adaptableness of administration and security. Likewise, standard access management techniques like role-based access control (RBAC) and attribute-based access control (ABAC) structures demonstrate their competence solving problems in tight conditions; the reception of access control frameworks will change the way consumers buy and use products and even the way they delegate their own privileges.

Another important aspect is granting reliable communications that are adaptable for the billions of gadgets ("things") connected in neighborhood, territorial, or worldwide systems. What is more, many of them are roaming or versatile gadgets and finding the location of and confirming the identity of a selected gadget is a remarkable issue for the IoT framework. This is only an example of the IoT challenges for governance, security, and privacy recognized in writing. The subsequent sections depict the difficulties distinguished by the accomplices of the AC05 joint ventures.

7.6 DIFFICULTIES FOR PRIVACY AND SECURITY IN IOT

The goal of this section is to distinguish the initial difficulties for the governance, security, and privacy in IoT recognized by the AC05 joint ventures and through discourses in IERC. Ethical views are in addition considered. Increasing weight is placed on ethics in IoT, and this idea is introduced in the section Ethics and Internet of Things; a number of difficulties are gleaned from the examination in this section.

7.6.1 SECURITY AND PROTECTION

This section provides a glance at structuring a security and protection system, which may address changes within the specific circumstance (e.g., an emergency or crisis) or conditions that hinder the gathering and handling of data from sensors. For example, in journalism, poor quality photos might prompt fake reports and hamper the safety and trust of the framework (e.g., level of notoriety).

The security and protection system has to offer greater accessibility and information on the particular state of affairs (e.g., access tips or access links assessed at time intervals). IoT allows improved applicability of the setting and further has to facilitate the coordination and reconciliation of many administrations, such as DiY (Do it Yourself) personalization flexibility, simplicity, and ease of use. An extra issue is that automated security protection innovations characterized for a selected setting might work incorrectly in an alternate (or unexpected) setting, thereby making people and their gadgets vulnerable.

Another aspect to be considered within the development of IoT with respect to essential administrations is the inescapability of mass-manufactured gadgets, that have increasing availability and re-configurability and consequently are helpless against malware of customary PCs. The fundamental issue is that these gadgets are progressively being incorporated into our regular day to day existence but they do not have the ability to actualize modern security protection packages like sure computing, or cryptography. As noticed, this setting presents difficulties of the ability to cope with the scope of the problem (billions of gadgets to ensure), harmonization, and consistency (various conventions and advancements).

7.7 CONCLUSION

IoT investigation is significant for modernization and this is confirmed by the IERC where there has been a tremendous effort in critical thinking to tackle the abundant demand for IoT supervision, security, and guard. Their work informs the key points of the IERC's IoT report. This report offers a blueprint for European investigation. Furthermore, it offers a structure to make it clear this IERC examination accumulates reports and fits them into one unit. This indicates any gaps where future examination is required or where it might be useful.

IoT superintendence is one of the biggest challenges. A good superintendence structure is fundamental to IoT's affluence over all edges, from design to activities to execution. IoT handles a broad range of discoveries, creating and propelling upwards to achieve openness, collaboration, accessibility, and scope for market-driven IoT advancement demands through a thorough organization framework that was up until recently non-existent. The dream of a self-sufficient organization, the progression of benchmarks agreed and maintained by experts across the globe, would give logical and supported action to shape up a trustworthy environment for multi-accomplice speculation and sponsorship. IoT security and assurance means keeping an eye on the needs of individuals, taking action for the future, making opportunities, and being astute to consumer demands while satisfying diverse needs across many areas. Execution, details, and costs are for the first half components that affect individuals who initiate trust. Though there have been vital progressions made and exercises to deal with accommodation, there are still many potential openings for "trust" framework, where additional investigation would be profitable.

Through the undertakings of the IERC, the IoT is on the right track. At any rate, the investigation continues, on account of a vision enabling specialists in IoT to flourish, upheld by astute businesses.

BIBLIOGRAPHY

Schindler, Helen Rebecca, Jonathan Cave, Neil Robinson, and Veronika Horvath. Petal Jean Hackett, Salil Gunashekar, Maarten Botterman, Simon Forge, Hans Graux. 2013. Europe's policy options for a dynamic and trustworthy development of the Internet of Things, RAND Europe. In: *Prepared for European Commission, DG Communications Networks, Content and Technology*, Europe.

EU IoT Task Force. 2012. *Final Report of the EU IOT Task Force on IOT Governance.* Brussels, November 14.

Weber, Rolf H. 2013. Internet of things – Governance quo vadis? *Computer Law and Security Review*, 29(4), 341–347. ISSN 0267-3649.

Roman, R., J. Zhou, and J. Lopez 2013. On the features and challenges of security and privacy in distributed internet of things, *Computer Networks*, 57(10), 2266–2279.

Miorandi, D., S. Sicari, F. D. Pellegrini, and I. Chlamtac 2012. Internet of things: Vision, applications and research challenges, *Ad Hoc Networks*, 10(7), 1497.

Conclusions of the Internet of Things public consultation. http://ec.europa.eu/digital-agenda/en/news/conclusions-internet-things-public-consultation. Last accessed 24 October 2013.

ENISA. 2013, November. Recommended cryptographic measures - securing personal data. http://www.enisa.europa.eu/activities/identity-and-trust/library/deliverables/recommended-cryptographic-measures-securing-personal-data.

ENISA. 2013, October. Algorithms, key sizes and parameters report. http://www.enisa.europa.eu/activities/identity-and-trust/library/deliverables/algorithms-key-sizes-and-parameters-report.

US White House. Big data: Seizing opportunities, preserving values. http://www.whitehouse.gov/sites/default/files/docs/big_data_privacy_report_may_1_2014.pd. 2014, May.

Paul, Subharthi, Jianli Pan, and Raj Jain 2011, January. Architectures for the future networks and the next generation Internet: A survey, *Computer Communications*, 34(1) 2–42.

Gama, Lionel Touseau and Didier Donsez 2012. Combining heterogeneous service technologies for building an Internet of Things middleware, *Computer Communications*. 35(4), 405–417. ISSN 0140-3664.

Gusmeroli, S., S. Piccione, and D. Rotondi 2013. A capability-based security approach to manage access control in the internet of things, *Mathematical and Computer Modelling*, 58(5), 1189.

Mehra, P. 2012. Context-aware computing: Beyond search and location-based services, *IEEE Internet Computing*, 16(2), 12–16.

Trappeniers, L., M. Roelands, M. Godon, J. Criel, and P. Dobbelaere. 2009, October 26–29. Towards abundant DiY service creativity successfully leveraging the internet-of-things in the city and at home. In: *Proceedings of the. 13th International Conference on Intelligence in Next Generation Networks*, Bordeaux, France, ICIN 2009.

Uckelman, D., M. Harrison, and F. Michahelles (eds.), 2011. *Architecting the Internet of Things.* Springer-Verlag: Berlin Heidelberg.

Marco, Conti, Sajal K. Das, Chatschik Bisdikian, Mohan Kumar, Lionel M. Ni, Andrea Passarella, George Roussos, Gerhard Tröster, Gene Tsudik, and Franco Zambonelli. 2011. Looking ahead in pervasive computing: Challenges and opportunities in the era of cyber–physical convergence, *Pervasive and Mobile Computing*, 8(1), 2–21, ISSN 1574-1192, doi:10.1016/j.pmcj.2011.10.001.

Cockton, Gilbert 2006. Designing worth is worth designing, *NordiCHI*, 14–18, 165–174.

Sørensen, Lene and Knud Erik Skouby (eds.), 2009, July. User scenarios 2020 – a worldwide wireless future. In: *OUTLOOK - Visions and Research Directions for the Wireless World, Wireless World Research Forum, No4*, Switzerland.

Ramachandran, A., L. Singh, E. Porter, and F. Nagle 2012. Exploring Re-identification risks in public domains. In: *Proceedings of the Tenth Annual International Conference on Privacy, Security and Trust (PST), 2012*, 35–42, Canada.

Bendlin, Rikke, Ivan Damgård, Claudio Orlandi, and Sarah Zakarias. 2011. Semi-homomorphic encryption and multiparty computation. In: *Proceedings of the 30th Annual International Conference on Theory and Applications of Cryptographic Techniques: Advances in Cryptology (EUROCRYPT'11)*, Kenneth G. Paterson (Ed.). Springer-Verlag: Berlin, Heidelberg, 169–188.

Gjøsteen, Kristian, George Petrides, and Asgeir Steine, 2012. Secure and anonymous network connection in mobile communications.

Issarny, V., N. Georgantas, S. Hachem, A. Zarras, P. Vassiliadist, M. Autili, M. Gerosa, and A. Hamida 2011. Service-oriented middleware for the Future Internet: State of the art and research directions, *Journal of Internet Services and Applications*, 79(1), 23–45. doi:10.1007/s13174-011-0021-3.

Christophe, B., M. Boussard, M. Lu, A. Pastor, and V. Toubiana 2011. The web of things vision: Things as a service and interaction patterns, *Bell Labs Technical Journal*, 16(1), 55–61. doi:10.1002/bltj.20485.

Yee, K. P. 2003, April 6. Secure interaction design and the principle of least authority. In: *Proceedings of the 21st International conference on Human Factors in Computing Systems – Workshop on HumanComputer Interaction and Security Systems*, Ft. Lauderdale, FL. CHI, Academic Medicine, New York, NY, USA.

Saltzer, J. H. and M. D. Schroeder 1975. The protection of information in computer systems. *Proceedings of the IEEE*, 63(9), 1278–1308. doi:10.1109/PROC.1975.9939.

Fenu, G. and G. Steri. 2009. Safe, fault tolerant and capture-resilient environmental parameters survey using WSNs. In: *Security and Privacy in Mobile Information and Communication Systems (First International ICST Conference, MOBISEC 2009, Turin, Italy, June 2009, Revised Selected Papers), Ser. Lecture Notes of the Institute for Computer Sciences, Social Informatics and Telecommunications Engineering (LNICST)*, A. U. Schmidt and S. Lian (Eds.). Springer: Berlin Heidelberg, 2009, vol. 17, 180–189. doi:10.1007/978-3-642-04434-2_16.

Cave, J. et al., 2011. *Does It Help or Hinder? Promotion of Innovation on the Internet and Citizens' Right to Privacy, Final Report*. European Parliament.

Ion, Mihaela, A. Danzi, H. Koshutanski, and L. Telesca. 2008, 26–29 February. A peer-to-peer multidimensional trust model for digital ecosystems. In: *Digital Ecosystems and Technologies, 2008. DEST 2008. 2nd IEEE International Conference on*, 469, Germany.

Benkler, Y. and H. Nissenbaum 2006. Commons-based peer production and virtue, *The Journal of Political Philosophhy*, 14(4), 394–419.

Benkler, Y., 2011. *The Penguin and the Leviathan*. Random House: New York, NY.

Copp, David, 2007. *The Oxford Handbook of Ethical Theory*. Oxford University Press: Oxford-New York.

Proper, Michel, 2011. *La philosophie du droit*. PUF: Paris.

Wiener, Norbert, 1950. *The Human Use of Human Beings*. Houghton Mifflin: Boston, MA.

Bynum, Terrell Ward and Simon Rogerson 1996. Global Information Ethics, *Science and Engineering Ethics*, 2(2), 131–247.

Luciano, Floridi and J. W. Sanders 2002. Mapping the foundationalist debate in computer ethics, *Ethics and Information Technology*, 4(1), 1–9.

Jeroen, Van den Hoven and John Weckert (eds.), 2008. In: *Information Technology and Moral Philosophy*. Cambridge University Press: Cambridge-New York, NY.

Moor James, H. 1985. What is computer ethics? *Metaphilosophy*, 16(4), 266–275.

Naess, Are 1973. The shallow and the deep, long-range ecology movement. A summary, *Inquiry*, 16(1–4), 95–100.

Cohen, Julie, 2012. *Configuring the Networked Self: Law, Code, and the Play of Everyday Practice*, Yale University Press: New Haven, NJ.

Benkler, Yochai and Helen Nissenbaum 2006. Commons-based peer production and virtue, *The Journal of Political Philosophy*, 14(4), 394–419.

Benkler, Yochai, 2011. *The Penguin and the Leviathan*. Crown Business: New York, NY.

Jasanoff, Sheila (ed.), 2011. *Reframing Rights. Bio-Constitutionalism in the Genetic Age.* MIT Press: Cambridge MA.

Jonas, Hans, 1984. *The Imperative of Responsibility.* University of Chicago Press: Chicago, IL (Frankfurt am Mein, 1979).

Dupuy, J. P. 2004, March 1–2. Complexity and uncertainty a prudential approach to nano-technology. European Commission. In: *A Preliminary Risk Analysis on the Basis of a Workshop Organized by the Health and Consumer Protection Directorate General of the European Commission*, Brussels.

Von Schomberg, R. 2012. Prospects for technology assessment in a framework of responsible research and innovation. In: *26 Technikfolgen abschätzen lehren: Bildungspotenziale transdisziplinärer Metho-den*, M. Dusseldorp and R. Beecroft (Eds.). Springer VS Verlag: Wiesbaden, 39–61.

Van den Hoven Jeroen et al., 2013. *Fact Sheet-Ethics Subgroup IoT -Version 4.0. Conclusions of the Internet of Things Public Consultation.* https://ec.europa.eu/digital-agenda/en/news/conclusions-internet-things-public-consultation.

Rebecca, Schindler Helen et al., 2013. *Europe's Policy Options for a Dynamic and Trustworthy Development of the Internet of Things.* SMART 2012/0053, RAND Corp., European Union, Europe.

Value Ageing' Project. Incorporating European fundamental values into ICT for ageing: A VITAL POLITICAL, ETHICAL, technological, and industrial challenge. Ref. Online. www.valueageing.eu.

Hannah, Arendt 1998. *The Human Condition.* University of Chicago Press: Chicago, IL (1958).

Hildebrandt, Mireille and Antoinette Rouvroy (eds.), 2011. *The Philosophy of Law Meets the Philosophy of Technology. Autonomic Computing and Transformations of Human Agency.* Routledge: London.

EDPS (European Data Protection Supervisor). 2010. *Opinion on Promoting Trust in the Information Society by Fostering Data Protection and Privacy (Opinion on Privacy by Design).* OJ C 280, 16.10.2010.

Nissenbaum, Helen 2011. From preemption to circumvention: If technology regulates, why do we need regulation (and vice-versa)?, *Berkeley Technology Law Journal*, 26(3), 1367–1386.

Pagallo, Ugo, 2012. Good Onlife governance: On law, spontaneous orders, and design, onlife project.   https://ec.europa.eu/digital-agenda/sites/digital-agenda/files/Contribution_Pagallo.pdf.

Snyder, Francis. 1993. Soft law and institutional practice in the European Community. European University Institute working paper, LAW no. 93/5.

Pasolini, G., D. Dardari, S. Severi, and G. Abreu. 2013, November 3–6. The effect of channel spatial correlation on physical layer security in multi-antenna scenarios. In: *Proceedings IEEE Fourty-Seventh Asilomar Conference on Signals, Systems and Computers*, Asilomar.
Cavoukian, 2011, January. Privacy by design: The 7 foundational principles, Revised Version.
Gusmeroli, S., S. Piccione, and D. Rotondi. 2012, September 2012. IoT@Work automation middleware system design and architecture. In: *Presented at 17 IEEE International Conference on Emerging Technology and Factory Automation (ETFA'12)*.
OAUTH Specification. http://oauth.net/2/.
OpenID 2.0. http://openid.net/specs/openid-authentication-2_0.html.
IoT6 D2.2. Distributed IPv6-based Security, Privacy, Authentication and QoS.
Jara, J. A, M. A. Zamora, and A. Skarmeta. 2012. Glowbal IP: An adaptive and transparent ipv6 integration in the Internet of Things, mobile information systems.
Jara, A. J., P. Lopez, D. Fernandez, J. F. Castillo, M. A. Zamora, and A. F. Skarmeta 2013. Mobile digcovery: Discovering and interacting with the world through the internet of things. *Personal and Ubiquitous Computing*. Springer-Verlag: London. doi:10.1007/s00779-013-0648-0.
IoT6 D3.1. Look-up/discovery, context-awareness, and resource/services directory.
IoT6 D2.3. Report on IPv6 based advanced features.
IoT6 D5.4. Intelligence distribution tests and evaluation report.
The HANDLE System. http://www.handle.net.
EPC information Services, Standard. http://www.gs1.org/gsmp/kc/epcglobal/epcis.
Neisse, Ricardo, Alexander Pretschner, and Valentina Di Giacomo 2013. A trustworthy usage control enforcement framework. *International Journal of Mobile Computing and Multimedia Communications (IJMCMC)*, 5(3), 3. doi:10.4018/jmcmc.2013070103.
Neisse, R. and J. Doerr. 2013, July 10–12. Model-based specification and refinement of usage control policies. In: *Privacy, Security and Trust (PST) Eleventh Annual International Conference on*, 169, 176. doi:10.1109/PST.2013.6596051.
Quartel, D. 1998. Action relations - basic design concepts for behaviour modelling and refinement. PhD Thesis University of Twente.
Fang, Lujun and Kristen LeFevre, 2010. Privacy wizards for social networking sites. In: *Proceedings of the 19th International Conference on World Wide Web (WWW '10)*. ACM: New York, NY.
Bagaa, M. et al., 2007. SEDAN: Secure and efficient protocol for data aggregation in wireless sensor networks. In: *Proceedings of IEEELCN*. IEEE, 1053–1060, Ireland.
Riggio, R. and S. Sicari. 2009, October, 1–6. Secure aggregation in hybrid mesh/sensor networks. In: *Ultra Modern Telecommunications & Workshops*, IEEE, USA.
Coen-Porisini, A. and S. Sicari. 2010. SeDAP: Secure data aggregation protocol in privacy aware wireless sensor networks. In: *Proceedings of the 2nd International Conference on Sensor Systems and Software*, Springer Verlag.
Mykletun, E., J. Girao, and D. Westhoff. 2006, September. Public key based cryp- toschemes for data concealment in wireless sensor networks. In: *Proceedings of IEEE ICC'06*, IEEE, 2288–2295.
Fragkiadakis, I. and E. Tragos Askoxylakis. 2013, September. Joint compressed-sensing and matrix-completion for efficient data collection in WSNs. In: *Proceedings of the IEEE CAMAD 2013*, Berlin, Germany.
Khan, Sarmad Ullah, Claudio Pastrone, Luciano Lavagno, and Maurizio A. Spirito 2012. An authentication and key establishment scheme for the IP-based wireless sensor networks, *Procedia Computer Science*, 10, 1039–1045.
Severi, S., G. Pasolini, D. Dardari, and G. Abreu. 2014, April 6–9. A secret key exchange scheme for Near field communication. In: *Proceedings of the IEEE Wireless Communications and Networking Conference, (WCNC, 2014)*, Europe.

Tragos, E. and V. Angelakis. 2013. Cognitive radio inspired M2M communications (invited paper). In: *IEEE Global Wireless Summit,* USA.
Golbeck, J. and J. Hendler. 2004. Accuracy of metrics for inferring trust and reputation. In: *Proceedings of the. 14th International Conference on Knowledge Engineering and Knowledge Management,* Berlin.
Golbeck, J. and J. Hendler. 2004. Inferring reputation on the semantic web. In: *Proceedings of the. 13th International World Wide Web Conference,* New York.
FaceBook-Connect API. http://developers.facebook.com/docs/guides/web.
Liberty framework. http://www.projectliberty.org/specs/ and Kantara http://kantarainitiative.org/.
Windows/Microsoft CardSpace. http://msdn.microsoft.com/en-us/library/aa480189.aspx.
ANTS, Gemalto. 2011, December. Oberthur technologies, and Safran Morpho, technical report: Restricted identification through access to e-services with privacy preserving credentials, France ANTS.
Gemalto, Oberthur. 2011, December. Technologies, and Safran Morpho, technical report: Privacy protocol semantic, criteria list and privacy-preserving credentials format, France ANTS.
Camenisch, Jan, Simone Fischer-Hübner, and Kai Rannenberg (eds.), 2011. *Privacy and Identity Management for Life.* Springer. ISBN 978-3-642-20316-9, Berlin.
Gurgen, Levent, Ozan Gunalp, Yazid Benazzouz, and Mathieu Gallissot. 2013, March 18–22. Self-aware cyber-physical systems and applications in smart buildings and cities. In: *Design, Automation & Test in Europe Conference & Exhibition (DATE),* 1149, 1154, Europe.
Dingledine, R., N. Mathewson, and P. Syverson. 2004, August. Tor: The second-generation onion router. In: *Proceedings of the 13th USENIX Security Symposium.* (Online). cites eer.ist.psu.edu/dingledine04tor.html.
Timpanaro, J. P., I. Chrisment, and O. Festor 2011, December. Monitoring the I2P network. Research Report RR - 7844, INRIA.
Clarke, Ian, Theodore W. Hong, Scott G. Miller, Oskar Sandberg, and Brandon Wiley 2002. Protecting freedom of information with Freenet. In: *IEEE Internet Computing,* Pennsylvania.
Bennett, K. and C. Grothoff 2003. Gap - Practical anonymous networking. In: *Designing Privacy Enhancing Technologies.* Springer-Verlag, 141–160, Berlin.
Rivest, R. and David L. Chaum 1981. Untraceable electronic mail, return addresses, and digital pseudonyms, *Communications of the ACM,* 24(2), 84–88.
Paquin, C. and G. Thompson, 2010. *U-Prove CTP White Paper.* Microsoft Corporation.
Chaum, D. 1983. Blind signatures for untraceable payments, *Advances in Cryptology - Proceedings of Crypto,* 82, 199–203.
Idemix, 2002. Camenisch & Van Herreweghen, design and implementation of the Idemix anonymous credential system.
Chaum Eugene and van Heyst 1991. "Group signatures". Advances in cryptology — EUROCRYPT '91, *Lecture Notes in Computer Science,* 547, 257–265.
Camenisch, Jan and Anna Lysyanskaya. 2001. Efficient non-transferable anonymous multi-show credential system with optional anonymity revocation. In: *Proceedings of Advances in Cryptology - Eurocrypt,* Zurich.
Lee, K. and M. Winslett Seamons and T. Yu. 2007. Automated trust negotiation in open systems secure data management in decentralized systems, *Advances in Information Security,* 33(Part III), 217–258.
Yajun, G. and W. Yulin. 2007, September 21–25. Establishing trust relationship in mobile Ad-Hoc network. wireless communications, networking and mobile computing, 2007. WiCom 2007. In: *International Conference,* 1562–1564, United States.

Devadas, S., E. Suh, S. Paral, R. Sowell, T. Ziola, and V. Khandelwal. 2008, April 16–17. Design and implementation of PUF-based "Unclonable" RFID ICs for anti-counterfeiting and security applications. In: *RFID, 2008 IEEE International Conference on*, 58, 64, Switzerland.

Internet of Things–Architecture IoT-A. Deliverable D1.5– Final architectural reference model for the IoT.

DG CONNECT trust and security. https://ec.europa.eu/digital-agenda/en/telecoms-and-internet/trust-security.

SysSec Red Book is a Roadmap in the area of Systems Security. http://www.red-book.eu/m/documents/syssec_red_book.pdf.

REGULATION of the European Parliament and of the council on the protection of individuals with regard to the processing of personal data and on the free movement of such data (General Data Protection Regulation) COM(2012) 11 final.

ETSI TErms and Definitions Database Interactive (TEDDI). http://webapp.etsi.org/Teddi/.

Directive 95/46/EC of the European Parliament and of the Council of 24 October. 1995 On the protection of individuals with regard to the processing of personal data and on the free movement of such data.

ITU. 2005. ITU internet reports 2005: The Internet of Things, executive summary, International Telecommunication Union.

Van Blarkom, G. W., J. J. Borking, and J. G. E. Olk, 2003. *Handbook of Privacy and Privacy-Enhancing Technologies: The Case of Intelligent Software Agents,* College bescherming persoonsgegevens, Netherlands.

8 Recent Trends of IoT and Big Data in Research Problem-Solving

Pham Thi Viet Huong and Tran Anh Vu

CONTENTS

8.1	IoT Big Data Exploration, Storage, Processing, and Analytics	136
	8.1.1 Data Exploration	137
	8.1.2 Data Processing	138
	8.1.2.1 Batch-Based Processing Technologies	138
	8.1.2.2 Technologies Based on Stream Processing	140
	8.1.2.3 Interactive Analysis	140
	8.1.3 IoT Big Data Analytics Methods	140
	8.1.3.1 Classification	141
	8.1.3.2 Clustering	142
	8.1.3.3 Association Rule	143
8.2	IoT Security	144
	8.2.1 Authentication Technologies in IoT	145
	8.2.2 Encryption and Key Management	146
	8.2.3 Protocols	147
	8.2.4 Challenges and Potential Future Trends in IoT Security	148
	8.2.4.1 Object Identification	148
	8.2.4.2 Authentication and Authorization	148
	8.2.4.3 Privacy	148
	8.2.4.4 Software Vulnerability Analysis	149
	8.2.4.5 Malware in IoT	149
8.3	The Internet of Things and Social Networks	149
	8.3.1 Basic Understanding of the Social Internet of Things (SIoT)	149
	8.3.2 SIoT Research Trends	150
	8.3.2.1 SIoT Paradigm	151
	8.3.2.2 SIoT Architecture	152
	8.3.2.3 SIoT Applications and Trends	152
8.4	Smart Cities and Applications	153
	8.4.1 IoT Technologies for Smart Cities	154
	8.4.1.1 Radio-Frequency Identification (RFID)	154
	8.4.1.2 Wireless Sensor Networks (WSN)	154
	8.4.1.3 Addressing	154
	8.4.1.4 Middleware	154

8.4.2 IoT Actual Trends for Smart Cities 155
    8.4.2.1 Smart Homes 155
    8.4.2.2 Smart Parking Lots 155
    8.4.2.3 Vehicular Traffic 156
    8.4.2.4 Environmental Pollution 156
    8.4.2.5 Surveillance Systems 156
    8.4.2.6 Smart Energy and Smart Grids 156
References 157

8.1 IOT BIG DATA EXPLORATION, STORAGE, PROCESSING, AND ANALYTICS

Big data is made up of data sets that are large and have great variety and velocity. Big Data has three main characteristics: volume, variety, and velocity, or the three Vs for short [1]. The *volume* of data is represented by its size, while *velocity* is the rate at which data is newly created. *Variety* means the different types of data and the various techniques used to examine the data. More recently, with the blossoming of the data era, the three Vs model is no longer enough to describe the data, and IBM has added a fourth V: veracity [2]. *Veracity* concerns biases, commotion, instability, and anomaly in data; hence it represents how accurate or truthful a data set may be.

There is a strong connection between Big data and IoT. Big Data processing is developed as the main IoT idea for improving decision making. The foremost potential feature of IoT is the capability of revealing knowledge about things in the networks. Hence, there is a great demand for the incorporation of Big Data in IoT applications. IoT and Big data have been already been used widely in information technology (IT) and business in recent times. Although the growth of Big Data is already slowing down, IoT and Big Data are dependent on each other and should be mutually developed. In other words, the implementation of IoT empowers both the quantity of data and the variety of data categories. Hence, IoT creates more chances for the evolvement of Big Data processing and analytics. Furthermore, the development of data diversity in the era of connected everything motivates the investigation of the operation of things in the network. The success of Big Data relies mostly on the ability to make sense of and generate useful information about connected things from the data. Given continuously changing data, in order to draw a meaningful conclusion, the process of merging Big Data with the administration of IoT can be categorized into three steps. The first step is data exploration, which manages different IoT data sources. IoT data sources are formed when things in the network communicate with one another. This data can be stored on the cloud. The obtained data are seen to have big volume, velocity, variety, and veracity, and hence are called Big Data. The second step applies Big Data analytics methods to organize the huge data available in the storage space into manageable data sets. The third step is the entire process of data analytics. Normally we learn from training data to see possible hidden information. After that, analytic tools are used in the testing data to confirm the conclusion given by the training data set. For example, the Big—Data, Analytics, and Decisions (B-DAD) framework has been presented in [1]. In this framework, Big Data analytics is utilized to make the right decision. All phases of the process are revealed in the model, including storage, management, processing, analytics, visualization, and evaluation. The following section furthers details each step.

8.1.1 Data Exploration

Data exploration is the first step in data analysis, which analyzes the main features of a data set, such as the size, accuracy, missing values, and initial information in the data. With the advances in technologies and the internet, a vast amount of data has become easily accessible for use in decision making. Data exploration, which is a time-consuming task, is the initial process of creating useful information from the insights. The continuous and rapid change in the type and variety of data requires new data exploration and analytics. In recent years, researchers have concentrated on creating new ways of exploring Big Data or have tried to modify the existing exploration techniques to make them fit the current trend of IoT.

Data exploration in the era of Big Data normally requires dividing the Big Data into several small, manageable data sets. These smaller data sets allow users to easily plot, view, and interpret the information from the insights. However, in recent literature, most researchers only use a random sampling with regular sized data sets. The study in [3] proposed that sampling techniques in data exploration can help generate insights by focusing on different aspects of the data without compromising data quality. Hence, the demand for different techniques of sampling would be promising for the future development of IoT and Big Data.

Several new Big Data explorations have been deployed in IoT Big Data. A new Big Data exploration called topological data analysis (TDA) is described in [4]. The purpose is to take advantage of topology, which is the mathematical study of shapes, to capture, represent, and inspect high dimensional real-world data sets. In recent years, there has been a boom in the generation of data in every sector such as biology, analytical chemistry, etc., in which the sample data is used to build simplexes, generalize intervals, and then stack together to form a frame approximation of the manifold. The frame and the manifold reveal the form or structure of the data. And the primary purpose of TDA is to measure the data's shape and its representation. The TDA has three key properties which make it suitable for Big Data: coordinate invariance, deformation invariance, and compression. The meaning of coordinate invariance is that the topological construction depends on the distance function that describes the shapes, other than a coordinates system. The second property, deformation invariance, means that the topology remains the same when a geometric shape is changed. The third property, compression, means that the TDA can illustrate all relationships in the data in a compressed form, one that is much simpler compared with the original one. Of course, in this case, we need to sacrifice a little bit of detail. Hence, TDA can handle continuously changing Big Data over time, which makes it a potential applicable method in the era of IoT.

Additionally, geospatial data handling methods are presented in [5]. In this research, geospatial Big Data methodology and theory as well as the major problems, issues, and challenges are reviewed and examined. In the circumstances of the Web of Data (WoD), following the abundance of linked data, several efforts have offered tools and techniques for exploration and visualization in many different domains.

However, most of these approaches are not successful in taking into account problems related to performance and scalability. Handing large data sets is crucial in the Big Data era. Therefore, [6] summarizes some possible directions for the future. Approximation techniques such as sampling and aggregation have been commonly used in systems from database and information visualization communities. Systems should be integrated with disk structures, retrieving data dynamic during runtime. Moreover, considering users' perspective, beyond visualization recommendations, modern WoD systems should provide more sophisticated mechanisms that capture users' preferences and assist them throughout large data exploration and analysis tasks. In [7], Big Data is closely connected to so-called datafication, the exploration of raw data in various aspects, which is considered an effort to solve complications and reduce instability. Datafication concerns the tools, technologies, and processes used to transform an organization into a data-driven business. The huge amount of everyday-life information is gathered and transformed into computerized, machine-readable data. When the data is digitalized, algorithms can be used to understand the hidden information.

8.1.2 Data Processing

Recently, the amount of obtained data from different sources has increased rapidly in many different ways. Therefore, the vital challenge in IoT Big Data is that there is not enough storage space to store the data. With that reality, methods for data storage and ways to extend the speed of input and output are urgently needed. In such cases, data access must be the primary concern for information discovery [8]. In the past, analysts use hard disk drives to store data, but the speed of the input/output is slow. To eliminate this constraint, the notion of solid-state drive (SSD) and phrase change memory (PCM) was presented. Nevertheless, current storage methods are still not able to store Big Data sets [8]. Hence, additional data reduction, data selection, and feature selection are required before we can apply techniques to understand the hidden information in the data. A promising research trend is expected to be in tools for processing Big Data sets. The data comes from different sources and needs to be stored by different mediums. The unorganized data is collected directly from traditional web-enabled databases [9]. Enormous analytics makes a difference in creating a commerce platform and increases competitiveness among companies by extracting important information from the obtained data. Additionally, it provides an efficient service; Big Data deployment will perform rapid analytics to enable organizations to get information quickly, make in-time decisions, and collaborate with other people and devices. In addition, this trend is even more important as existing algorithms are not always able to deal with Big Data. In recent years, numerous tools have been investigated for processing Big Data. Most of the available tools focus on batch processing, stream processing, and interactive analysis.

8.1.2.1 Batch-Based Processing Technologies

8.1.2.1.1 Apache Hadoop and MapReduce

Apache Hadoop is a well-known tool for handling large data sets. Currently, a lot of companies are utilizing Apache Hadoop technology in their business, including

Recent Trends of IoT and Big Data

SwiftKey [10], Nokia [11], and Alacer [12], to name but a few. Apache Hadoop comprises of Hadoop kernel, MapReduce, and Hadoop distributed file system (HDFS). Hadoop uses the MapReduce programming model, which is based on the method of division and conquest issued to process large amounts of data. The master nodes and worker nodes work together in Hadoop. The master node is responsible for dividing tasks among worker nodes. When all worker nodes complete the works, they return small parts to the master nodes. Then the master nodes combine every output in reduced steps. Hadoop has many advantages; for example, it can process distributed data, perform tasks independently, and handle partial errors easily [13]. However, Hadoop still has some disadvantages, such as a restrictive programming model. It has only a single master node, and the distribution and configuration of the nodes is not obvious [13]. Despite its disadvantages, however, Hadoop is still a powerful software framework for Big Data problem solving.

8.1.2.1.2 Dryad

Dryad, DryadLINQ, and Distributed Storage Catalog (DSC) are a set of advance technologies designed to maintain the intensive data processing in the Windows platform's application [14]. Dryad uses a dataflow graph for actualizing parallel and distributed programs. The Dryad graph is called a directed acyclic graph (DAG), comprising vertices and channels. A vertex illustrates the program to process the data in a predefined way. The edges represent the channels that transmit data from vertices to vertices. In Dryad, the nodes are divided into a cluster, and a user employs the resources of these clusters to execute the program in a disseminated way. The advantage of Dryad is that it does not require knowledge about concurrent programming. A higher-level programming language and compiler for Dryad is DryadLINQ, which is able to naturally compile the Language-Integrated Query (LINQ) programs composed in .NET into optimal computational steps. The Distributed Storage Catalog (DSC) works together with the NTFS (New Technology File System) to supply data administration functionalities, such as data collection, storage, replication and load balancing in cluster.

A typical Dryad application works by executing a computational directed graph that consolidates nodes and channels. Subsequently, Dryad offers great functions: making a work graph, planning equipment for the accessible processes, tracking error processing within the cluster, collecting execution parameters, visualizing the job, and automatically updating the work graph. The study in [14] assessed Dryad's running time and communication costs in practical settings. An execution model which shows Dryad via parallel matrix multiplication (PMM) is provided. Exploratory analytics are performed to confirm the accuracy of the examination model on Windows cluster with up to 400 cores, Azure with up to 100 entities, and Linux cluster with up to 100 nodes. The ultimate outcomes reveal that the analytic model gives a precise estimation accurate to within 5% of the measured results.

8.1.2.1.3 Jaspersoft

Jaspersoft is an open-source software package which can generate reports from database columns. Jaspersoft is an expansive information analysis configuration that can expand and visualize quick data on prevalent storage platforms, including

MangoDB, Cassandra, and Redis. The advantage of Jaspersoft is that it is conceivable to rapidly investigate Big Data with no requirement on extraction, transition, and loading (ETL). Furthermore, it is also able to build efficient hypertext markup language (HTML) reports and dashboards, which can be shared among the network directly from a large data store without ETL.

8.1.2.2 Technologies Based on Stream Processing

8.1.2.2.1 Storm

Storm is specially designed for expansive streaming data, which is a distributed and fault-tolerant computing framework. Storm is specially provided to handle real-time data, while Hadoop is primarily used for batch processing. Moreover, another advantage of Storm is its simple set up and operation for competitive performances. Storm clusters are comparable to Hadoop clusters, which incorporate master nodes and worker nodes. Users run diverse structures for different types of tasks. The master nodes and worker nodes work as nimbus and supervisor. The two roles have comparative functionalities according to the job tracker and task tracker of the MapReduce technology. The nimbus disseminates code on storm clusters, plans and assigns tasks to worker nodes, and monitors the whole system. The supervisor adheres to tasks assigned by the nimbus. Moreover, the supervisor starts and ends the process as required based on the instructions of the nimbus. The entire computing technology is partitioned and then distributed to a number of worker processes, and each worker process performs a part of the topology.

8.1.2.2.2 Splunk

Splunk is a smart real-time configuration designed for large-scale data mining. It integrates cloud technologies and Big Data. Splunk's main purpose is to help users discover, track, and analyze data produced by their machines via the Internet. The outcomes are presented in a visual way, such as through graphs, reports, and warnings. Splunk differs from other stream processing tools, including structured and unstructured data indexing, real-time search, analysis of results, and boards. The significant goal of Splunk is to generate parameters for multiple applications, troubleshoot diagnostics for information technology framework and systems, and provide smart support for organization operations.

8.1.2.3 Interactive Analysis

A typical interactive data analysis is Apache drill, which has the adaptability to boost a diversity of query languages, data formats, and data sources. It is especially outlined for nested data mining. In addition, it is able to scale up on 10,000 or more servers and is capable of handling petabytes of data and trillions of records in seconds. Drill uses the Hadoop Distributed File System (HDFS) for data storage and MapReduce for batch analysis.

8.1.3 IoT Big Data Analytics Methods

The development of social networks and the IoT, as well as the vast amount of information stored by companies, are creating rapid growth in the amount of data generated. Therefore, data analysis techniques must be enhanced to process this huge array of information to disclose hidden useful information. The ability to investigate huge amounts of data can help a company process significant operation and business information. Accordingly, the primary goal of Big Data analytics is to immediately extract knowledgeable information to create accuracy predictions, recognize recent patterns, discover hidden information, and, consequently, make decisions [15]. Hence, technologies and tools are required in Big Data to transform a large amount of structured, unstructured, or semi-structured data into a more manageable analytical data. After the data is analyzed, the information extracted from the data is visualized in tables, graphs, and spatial charts for efficient and timely decision making. Therefore, Big Data analysis is a significant challenge for a variety of applications due to its complexity and the adaptability of the applied scenarios.

Data mining plays a major role in analytics, and most Big Data methods are created based on data mining techniques in accordance with a specific circumstance. The extracted information from the analytics of available data plays an important part in assessing and choosing a suitable method for decision making. In this section, numerous Big Data analytics are presented in classification, clustering, association rule mining, and prediction categories.

8.1.3.1 Classification

Classification is a supervised learning method that utilizes previous knowledge as training data to divide objects into groups [16]. In classification, a predefined category is assigned to an object, and therefore, the predicted outcome of a class for an object is obtained. Finding unknown or hidden patterns is more challenging for Big Data in IoT. Moreover, extracting valuable information from large data sets to improve decision making is a critical task. An example of a classification method is Bayesian network, which provides a graphical and interpretable representation of hidden information extracted from the data. Since the output of the Bayesian network classifier is a probabilistic model, decision theory is naturally applied to address cost-sensitive issues, thus providing a reliable measure on the predicted label. Bayesian networks are efficient for analyzing complex data structures that are disclosed through Big Data instead of traditional structured data formats. Moreover, they are suitable for more complex classification problems in any kind of domain such as discrete, continuous, and mixed data; new flows of streaming data; and so on. The nodes and edges in Bayesian network are represented as random variables and conditional probability, respectively [17].

Another widely known classification method applied in Big Data analytics is Support Vector Machine (SVM), which is a statistical method to visualize the distribution of the data and, from that, to divide data into groups for simpler analytics. SVM works as follows: If we are given a set of training examples, each record is marked as belonging to each of the categories, and the training model assigns new records each suitable category. The purpose of SVM is to create an optimal hyperplane to categorize new records. In SVM, all records are represented as points in space. SVM works by mapping these points to divide the categories by a gap that is as wide as possible. When performing the classification task, new records are mapped into that same space and are predicted to belong to a certain category based on which side of the gap they fall. In a two-dimensional space, the optimal hyperplane is a line that divides a plane into two parts. SVM can not only perform linear classification but it is also good at performing non-linear classification with the help of the kernel trick, which maps their input into high dimensional spaces. This feature makes SVM a critical technique in IoT Big Data. Some applications of the SVM technique are text classification [18], pattern matching [19], and health diagnostics [20].

Similarly, k-nearest neighbor (k-NN) is often used in search applications where you are looking for "similar" items, that is, when your aim is some form of "find items similar to this one." It is particularly implemented to find hidden information in a data set. The biggest use case of k-NN search might be the Recommender Systems [21]. If it is known that a user prefers a specific item, it is recommended to suggest similar items. To find similar items, the groups of users who like each item are compared. If a similar group of users like two different items, then the two different items are probably similar. This applies to recommending products, recommending media to consume, or even "recommending" advertisements to display to a user. Additionally, k-NN has been widely applied in other areas, for example, in anomaly detection [22], high-dimensional data [23], and scientific experiments [24].

8.1.3.2 Clustering

Clustering algorithms can be considered as an alternative super-powerful learning tool to accurately analyze huge volumes of data. Unlike classification, clustering is an unsupervised learning technique, in which generally there is nothing to illustrate how the data should be grouped together. In other words, it is a way to group similar data patterns in a predefined way. In clustering, the records within a group are as similar as possible, while the records between clusters are as dissimilar as possible. When the data is divided into a relatively small number of clusters, we can have a better understanding about the data's hidden information, although we need to compromise at the price of losing some detail.

8.1.3.2.1 Hierarchical-Based

The hierarchical clustering method combines several data clusters to form a hierarchical dendogram. Data is organized in a hierarchical manner depending on proximities, which are calculated by the intermediate nodes. In a hierarchical cluster, the data set is represented by a dendrogram, in which individual data is presented by leaf nodes. There are two ways of hierarchical clustering, which are agglomerative (bottom-up) and divisive (top-down). They work in opposite ways. In agglomerative clustering, each record is assigned as one cluster, and these clusters are recursively merged until the optimal number of clusters is reached. Division works in the opposite way. The original data is considered as the initial cluster, which is gradually divided into several smaller clusters as the hierarchy continues [25]. The process continues until a stopping criterion is reached (normally, the required number k of clusters). Instead of analyzing a vast amount of data, we just need to investigate the information revealed by clusters. However, the hierarchical method has

Recent Trends of IoT and Big Data 143

a major drawback, as when merging or splitting is performed, it cannot be undone. Some of the hierarchical clustering algorithms are BIRCH, CURE, ROCK, and Chameleon [26].

8.1.3.2.2 Partitioning-Based

In partitioning-based clustering, initial groups are assigned and reallocated in the direction of consolidation. In other words, partitioning algorithms divide data records into several partitions, where each partition represents a cluster. These clusters must meet the following requirements: each group must contain at least one object, and each object must belong to exactly one group. A typical partitioning-based algorithm is the K-means algorithm, in which a center is the average of all points and coordinates representing the arithmetic mean. All records are divided into K clusters in which each record is assigned to the cluster with the nearest mean. There are also many other partitioning algorithms, such as K-modes, PAM, CLARA, CLARANS, and FCM [26].

8.1.3.3 Association Rule

Association rule represents the frequent co-occurrence of items in a data set. Association rule mining has a number of applications, and the most widely used application is to discover sales correlations in transactional data—in other words, market analysis. It allows the prediction of the future operation of each organization based on the understanding of sales information, for example, changes in prices or changes in sales volume. The purpose of association rule techniques [27] is based on the frequency of occurrences to identify and create rules for quantitative data. Data processing is performed in two ways under association rules: priori-based algorithms and temporal sequence analysis. Priori-based algorithms have been used in MSPS [28] and LAPIN-SPAM [29], which show considerable effectiveness in mining sequential patterns. Temporal sequence analysis is less widely used. It serves the purpose of understanding the hidden patterns in continuous data.

Prediction is another critical task in IoT Big Data analysis. The purpose of predictive analytics is to use historical data (normally called the *training* data) to determine the behavior of components in the network. Regression algorithms are normally used to determine the relationships between independent (input) and dependent (output) variables. From the regression curve, prediction of the output can be carried out. There are several researches that have been pulled out. In [30], we can predict customer purchase and social media trends. Time series analysis is also used for the purpose of prediction. The advantage of time series analysis is that it can reduce the high dimensionality of Big Data, hence creating a better data visualization for better prediction. Research related to time series representation includes ARMA [31], bitmaps [32], and wavelet functions [33].

The major Big Data analytics strategies examined in this chapter are broadly applied in various areas of application, such as industry, trade, catastrophe organization, and healthcare. Methods of data analysis based on clustering and association rules are appropriate to industry and e-governance and are adopted in healthcare and e-commerce. Predictive analytics are valuable for disaster and market forecasts, while time series analysis is utilized in a number of applications such as disaster prediction, imaging restoration, speech recognition, and social network analytics.

8.2 IOT SECURITY

The IoT is growing quickly due to the multiplication of communication innovation, the accessibility of devices, and computational frameworks. For the most part, the core group of "things" themselves has developed to cover different sorts of devices, from straightforward radio frequency identification (RFID) tags and remote wireless sensor devices to complex frameworks like connected cars, TVs, cameras, and so on. Subsequently, IoT security has become an area of concern in order to protect the equipment and the systems in an IoT framework. As the IoT keeps evolving over time, the security for IoT also evolves and changes according to the field it secures. The critical nature of the concept of security within the IoT is also advancing. This can be outlined in Figure 8.1, which shows the proportion of IoT articles that unequivocally mention the term "security" according to Google scholar (as of December 2017) [34]. The blue part shows the percentage of papers that relate to security, while the yellow part shows the percentage of papers that do not mention the term "security" at all. We can easily see not only that the percentage increases with time, but also that the number of papers is also growing at a great rate. This chapter concentrates on examining current trends in how IoT security evolves over time. On this basis, we can arrange and create more optimal security mechanisms to ensure our connected "things" in the future.

Figure 8.2 shows the annual patterns dissemination of IoT security articles in the period 2011–2016 [35]. This indicates that authentication is the most source of recent IoT security advances, which is the subject of the most prominent number of articles especially in recent years (2015, 2016). Recent development in the number of articles on all the other innovations is additionally clearly shown.

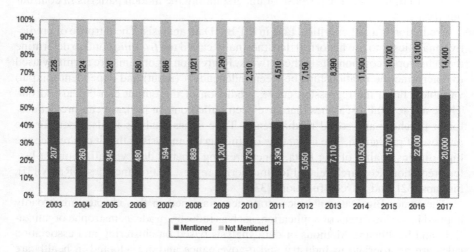

FIGURE 8.1 IoT research articles that explicitly mention the term "security" [9].

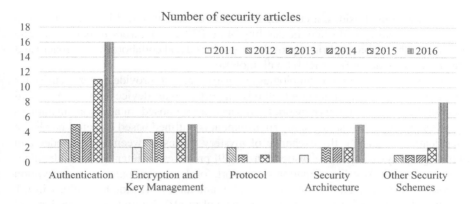

FIGURE 8.2 Distribution of IoT security articles by year (2011–2016) [35].

8.2.1 AUTHENTICATION TECHNOLOGIES IN IoT

Authentication, in basic terms, is the method or activity of providing or showing something to be true or valid. Within IoT, authentication advances are utilized to recognize the data of IoT substances such as devices and users to avoid malicious attackers from expanding unauthorized access. To guarantee secure communication among devices, analysts have created different authentication strategies. As seen in Figure 8.2, 43% of the chosen papers are on authentication related to the IoT environment. This reveals that authentication is the current primary stream of investigation in IoT and Big Data.

Numerous authentication strategies have been created for IoT and Big Data. RFID is seen as one of the prerequisites for the execution of IoT as it can unravel the unique identification issues of IoT. Additionally, RFID tags are able to store and wirelessly transmit information and offer the opportunity to automatically recognize and track objects. Ordinarily, there are three fundamental components in an RFID framework: tags, readers, and a back-end server. The identification protocol works as follows: RFID readers send a request to RFID tags, which are attached to objects or individuals. The tags take the request and send a response which contains an identification data back to the reader. The reader forwards the tags' responses to the back-end server. Finally, the back-end server performs tag identification by taking care of the request and the response of the tag. The investigation in [36] proposed a destructive private RFID authentication protocol. This protocol can be utilized for large-scale RFID frameworks. It employs only one master key shared by all tags, which has consistent time identification, but no security is performed as soon as one tag is compromised. It is based on the utilization of physically unclonable functions (PUFs) using random physically variations. As far as we know, this study is the first protocol giving a higher security level without lookup property. A new RFID authentication scheme is created with anonymity, untraceability, and forward secrecy inside the partial distributed server model that gives only part of the authentication information among the servers [37]. In this scheme, the computation on the tag is affordable on the low-cost cryptographic tags, and its fabulous execution makes it appealing for

IoT applications. Besides, a customizable and versatile protocol is given in [38] to guarantee security and scale the sending of an RFID framework to boost IoT. The protocol is based on a hybrid approach (group-based and collaborative approach) and security check handoff for RFID with mobility.

Mutual authentication technologies are also created to provide better security for IoT than one-way authentication methods. In [39], an inter-device authentication and session-key distribution framework for devices are proposed. In addition, an authentication protocol for a smart home environment is outlined based on zero knowledge verification instead of the exchange of mystery values between sensor nodes and home gateways [40]. The investigation in [39] proposes an inter-device authentication and session-key dissemination framework for devices prepared with encryption modules. A mutual authentication innovation and anonymization technology in IoT security for utilization of Big Data is inspected in [41]. Mutual authentication and access control between connected devices can protect devices from external attacks and secure data communication between devices. In IoT devices, we need to make the most of the data. Hence, authentication technology is required not only to link the device and the person but also to connect the sensor ID to the personal ID.

Multi-factor authentication strategies, which require more than one method of authentication to confirm the user's identity for a login or other transaction, give additional security to communications between devices and systems. Multi-factor authentication may comprise a MAC address, ID, password of mobile devices and MAC address, or Ipv6 address of controlled objects [42]. These variables are fingerprinted and changed into a unique value, which plays a critical part in identifying the subjects for the proper access control in the IoT environment. Another possibility is a two-factor authentication protocol based on two-way authentication for RFID systems [43].

Another authentication technology, called a lightweight authentication, is designed to optimize the particular constraints of each system. It is essential for solving security and privacy problems. For E-health applications, a safe group-based lightweight authentication scenario is proposed in [44]. The model offers an extensive security level against more than one type of attack, such as impersonation attacks, man in the middle attacks, and unknown key sharing attacks in healthcare IoT based applications. Another lightweight mutual authentication for smart IoT and wearable devices is mentioned in [45]. Since smart and wearable devices interact with a variety of sensors and they must be connected to the Internet, they provide a wide range of IoT services to clients. Hence, a proper security scheme is required for safe service operation. [46] provided a lightweight RFID mutual authentication protocol with a cache within the reader, lessening computational and transmission costs.

8.2.2 Encryption and Key Management

Encryption is the method of encoding a message so that it can be read only by the sender and the authorized recipient. To read an encrypted record, you must have a secret key that empowers you to decrypt it. The essential reason for encryption is to secure the privacy of computerized information stored on computers or transmitted via the web or other computer systems. Subsequently, data encryption is critical for

Recent Trends of IoT and Big Data 147

information privacy, which is an IoT security requirement [47]. However, encryption innovations cannot secure the source and destination of the data [48]. To fill this gap, key administration algorithms have been examined to permit only authorized users to access the system and avoid malicious attacks.

Recently, companies have been utilizing IoT devices to gather real-time and continuous data and make better business choices to improve their returns and customer satisfaction. Collected information needs to be prepared and transferred in an appropriate format to be stored in a storage system. This information grows exponentially, so cloud storage for IoT data is essential. For reliable and safe communication within the cloud-based IoT framework, research is being carried out into [49] several access control arrangements and cryptography concepts for companies to store the IoT data on the cloud safely.

Security in the smart home is also an interesting topic. The study in [50] is aimed at enhancing security in smart home systems. Smart home systems utilize encryption and hash algorithms, which can securely send messages to each other. The devices involved can be the thermostat, air conditioners, lighting systems, and so on.

With respect to IoT in the healthcare division, the security and integrity of the medical data has created huge challenges. A hybrid security model is proposed for securing the diagnostic text data in medical images [51]. The model is created through joining either a 2-D discrete wavelet transform 1 level or 2-D discrete wavelet transform 2 level steganography procedure with a hybrid encryption scheme. Thus, the proposed model demonstrated its ability to hide the private patient's information in a transmitted cover image with high imperceptibility, capacity, and negligible deterioration in the received image. Moreover, [52] has considered the proposed method in a hybrid encryption algorithm, which is utilized as a portion of IoT. It is the combination of grasshopper calculation (GO) and particle swarm optimization (PSO). The proposed calculation, which is hybrid swarm-based optimization, took less time for both the encryption and decryption processes. In [53], a new cryptography strategy, named elliptic Galois, is utilized to encrypt confidential data that came from distinctive sources. Then, a Matrix XOR encoding method is used to implant the encrypted data into a low-complexity image. Inside the image, an Adaptive Firefly optimization calculation is used to optimize choice of cover blocks. Finally, the information that is hidden in the image is recovered and is then decrypted.

8.2.3 Protocols

According to Figure 8.2, eight papers have proposed new novel security protocols to guarantee more secure communications in IoT networks. For example, [54] explored security protocols given by communication advances utilized in IoT such as RFID, Bluetooth, Wireless network, and ZigBee. The research in [55] outlined a secure framework for IoT devices using the constrained application protocol (CoAP) over the datagram transport layer security (DTLS) protocol. This scheme helped enable secure communication between resource constrained things. [56] provided a secure communication model for CoAP, a lightweight communication model planned for smart-object networks. In addition, [57] proposed a communication protocol based on hash lock, session keys, random numbers, and security keys designed to be secure against hacking from intruders. In [58], a protocol for multi-hop communications between device-to-device (D2D)-enabled terminals is designed and tested.

8.2.4 Challenges and Potential Future Trends in IoT Security

8.2.4.1 Object Identification

Object identification is critical in IoT security. The biggest challenge in object identification is to guarantee exactly the records used in the naming design. In spite of the fact that the Domain Name System (DNS) offers name interpretation services to Internet users, it is still an unreliable naming framework [59]. The reason is that the DNS is unprotected against different attacks, for example, DNS cache harming attacks and man-in-the-middle (MITM) attacks. This harming attack infuses fake DNS records into victims' cache and specifically accommodates the determination mapping between naming framework and addressing framework. Hence, without genuine protection of the records, the whole naming framework is defenseless. The expansion of DNS, which is Domain Name Service Security Extension (DNSSEC, IETFRFC4033), can ensure the integrity and authenticity of a resource record (RR). DNSSEC can overcome the limitations of naming services; however, it is challenging to implement DNSSEC in IoT [59]. Hence, a new naming service is expected in the near future of IoT security.

8.2.4.2 Authentication and Authorization

Regarding authentication and authorization, the demand for a global root certificate authority (CA) anticipates numerous theoretically attainable plans before actually being deployed. The CAs are unquestionably small data files that include identity credentials to assist websites, individuals, and devices represent their authentic online identity. CAs play an imperative part in how the Internet works and how valid transactions can take place. Despite the reality that public-key cryptosystems are dominant in developing authentication frameworks, global root CAs are still necessary. If there is no CA, it is exceptionally difficult to create an authentication framework for IoT systems. Moreover, it may be impractical to allocate a certificate to an object in IoT since the total number of objects is regularly intensive.

8.2.4.3 Privacy

The challenges in preserving privacy in IoT can be split into two types: data collection policy and data anonymization. Data collection policy depicts the course of action inside the data collection step, where it powers the sorting of realistic data and the control of the information. The sort and the sum of data to be collected is limited in only the data collection step. Security conservation can be guaranteed by confining the collection and storage of private information [59]. Another challenge is data anonymization. This is a method of either encrypting or concealing information, which is the process of removing personal identification information from data sets so that the individuals related to the portrayed information remain anonymous. Given the diversity of the things, numerous cryptographic plans can be embraced. An example of a cryptographic scheme is lightweight, which is more appropriate to devices that are resource limited. Otherwise, concealment of information can be achieved by using data encryption in cases when the obtained data are resistant to data investigation. However, information ought to be shared among things in the IoT network; in this manner, calculation on encrypted data is another challenge for data anonymization.

8.2.4.4 Software Vulnerability Analysis

The revelation of vulnerability before the release of a product requires dynamic analysis. In spite of resource limitations, dynamic investigation may be wasteful to convey in an IoT device. Subsequently, simulation is required to carry out dynamic analysis. Simulation involves recreating the behavior of devices in a server. However, the semantic gap between real device and simulated framework is an imperative issue which needs to be deal with. The diverse equipment in a device such as a GPS makes it indeed more troublesome to close the gap. The discrepancy between device and emulated framework is hard to avoid. Numerous examination methods are profoundly dependent on the system, such as taint analysis and symbolic execution. An investigation system must be sufficiently adaptable to embrace diverse systems. Nevertheless, we need to consider management and policies, rather than simply technical issues.

8.2.4.5 Malware in IoT

As mentioned above, the risk of viruses or malware is significant because of the constrained assets of IoT devices. Besides, traditional security mechanisms to oppose malware can be impractical when moving straight from the common ×86 architecture platforms to the smart IoT platform. For example, people normally use antivirus software as one of the foremost viable security tools for detecting known real-time malware. Nevertheless, the computing power of IoT devices is relatively small, which is different from the ×86-architectured PC. Thus, the real-time scanning functionality of antivirus software may result in the consequence of unreasonably expensive overhead for IoT devices. In the meantime, malware creators who take into account the computing power problem of IoT will adapt their malware to target the isolated downloader and the main body. The downloader has easy access to devices, and so if the downloader contains malware, it is ready to contaminate any IoT devices that have a small program body. The divergence of hardware configuration is also a good source of malware.

8.3 THE INTERNET OF THINGS AND SOCIAL NETWORKS

8.3.1 Basic Understanding of the Social Internet of Things (SIoT)

The connection between the IoT and social networks (SN) empowers the association of individuals to the *ubiquitous computing* universe. The term *ubiquitous computing* refers to computing power that can be accessed anytime and anywhere [60]. In the IoT system, data about the environment is provided by the IoT, and the SN brings the glue that permits interactions between people and devices [61]. IoT will open a variety of smart services and applications to take advantage of numerous opportunities that people and companies face in their everyday lives. Thus, it allows people and things be connected with anyone or anything, in any place, at any time. Given the convenience provided by the IoT, people will use smart services and applications to improve their quality of life. Subsequently, the IoT, where things are able to build up social connections with other objects, independently with respect to people, is called the Social Internet of Things (SIoT). The quality of experience (QoE) of applications depends on how the motivated demand from the human is fulfilled. In order to integrate the ubiquitous computing in humans' daily life with high QoE, the network between users and things needs to be improved. Hence, we take into account social networks of people and things for worldwide computing as an advance beyond IoT. The connections here not only mean the physical connections between humans and things but also the logical configurations of social communities related to people and things.

The term *community of interest* (COI) often relates to a set of communicating entities. Individuals in such COI share different resources such as contextual relationships, interests, and resources in both online and offline settings. Social connections and contextual information shared within a web community are considered as beneficial resources that encourage collaboration for mutual benefit. The SIoT extends this phenomenon to coordinate physical components and their data from the real world into the SN to achieve improved knowledge and modeling of users' real demands and targets. Hence, the SIoT brings up two fundamental benefits: improving the QoE and strengthening the collaboration within communities including people and things [61].

While IoT studies primarily address communication with the physical world by sensing through numerous distinctive devices, the SIoT diagram raises vital concerns about the reasons for and use of these services and applications. In order to improve the QoE, we need to enhance the network of connections between people and things. In human society, a person often functions as both a consumer and a producer to communicate with others. In the physical and online world, SN may give proactive (long term) or reactive (on demand) information on such things as needs, interests, locations, and collaboration characteristics of relationships among people. And these social aspects could be stored, processed, and analyzed to assist in improving the QoE. With respect to pervasiveness, people tend to access services anytime and anywhere using any type of device through any type of communication systems. Within the IoT context, pervasiveness implies the weaving of real, virtual, and physical things into the SIoT.

In SIoT, users can inadvertently join in the process of enhancing QoE through the things they normally use and share in their everyday life. In addition, things will cooperate with other different things to meet humans' targets. In this collaborative scheme, people and things are not isolated nodes in a network; their targets and demands will combine to shape the SIoT.

8.3.2 SIoT Research Trends

Although the concept underlying the SIoT is at an early stage of development, some studies have come up with solutions for connecting people with distributed sensors and embedded devices. The research in [62] reported a case of a socio-technical network, in which both humans and technical frameworks talk and exchange information to realize the vision of ubiquitous computing environment in the future. The IoT mechanism is said to be enriched with Twitter communication abilities to update and post information about the status of tasks and activities. Additionally, in [63], a remote sensor system is utilized to develop a method that can use Twitter to distribute sensory data and resources. In [64], a framework is designed to use SN to encourage the cooperation and association of the distributed SN. Currently, the mainstream in SIoT shows the following trends:

8.3.2.1 SIoT Paradigm

The SIoT paradigm is characterized as a biological system that permits individuals and smart devices to connect within a social network. In this section, key aspects that constitute the premise of SIoT are summed up. They are the social role, intelligence, socialized devices, and everything as a service.

8.3.2.1.1 Social Role

At its simplest, a social role can be characterized as a set of connected behaviors, rights, obligations, and commitments conceptualized by humans in a social circumstance. In the IoT concept, social role may be a set of users' activities to maintain the social structure and connection with physical devices in the networks. In [65], the social role is displayed by the prevalent online SN and their application programming interface to preserve a social structure and connections with smart objects. Similarly, in [66–68], the social role is outlined by bringing users into the IoT world to ensure the network navigability, and to produce better benefit.

8.3.2.1.2 Intelligence

In the literature, intelligence is shown to be a type of decision-making that promotes the use of services. Regarding the IoT, in [68], the notion of intelligence is specified as a fundamental component of the SIoT paradigm which is capable of the whole process, from starting and updating a network to terminating the links among things in IoT. As a distinctive approach, in [65], the intelligence is expressed in the form of dynamic object service discovery in which intelligent objects can automatically understand others' services.

8.3.2.1.3 Socialized Devices

The definition of socialized devices was introduced in the early stage of SIoT [65–68] as the means for smart things, embedded devices, and people to communicate with each other. Hence, it is considered the foremost fundamental components in the SIoT [61]. The research in [69] presented some concepts for devices that are providing Internet-friendly social capabilities and described a number of prototypes of devices that used the Internet to communicate with others. In [65], social devices can communicate with users through Web protocols in a SN.

8.3.2.1.4 Everything as a Service

The tendency to move SIoT functionalities into services and allow them to be found and integrated with different existing services has been considered a challenge in the study of IoT and Big Data. To take advantage of the convergence of the devices and the social roles, [70] proposed a platform that empowers individuals to share their Web-enabled devices so that others can utilize them. The structure is leveraged based on existing social networks and application programming interface. In addition, the service in [71] is outlined as the appropriateness of the social networks by coordinating autonomous and proactive things.

8.3.2.2 SIoT Architecture

SIoT architecture is also a promising research trend. Several research projects have attempted to investigate the architecture of the SIoT. For example, the research in [72] handled future IoT architecture in two angles: unit IoT and ubiquitous IoT. The model for future IoT is not only beneficial in understanding the connection between the IoT and its encompassing environment but also useful for the current improvement of the execution of IoT.

8.3.2.3 SIoT Applications and Trends

The potential advertised by SIoT can create a mass of applications and trends. As this concept is still new, there are gaps for researchers to mine. Some current trends are listed below.

8.3.2.3.1 Semantic Web Service

Web service is the core of the SIoT; therefore, web services' improvement makes SIoT more feasible. Suitable policies are mentioned for the foundation and administration of social connections among objects in such a way that a social network can be navigable [66]. An IoT architecture is provided, which incorporates the necessary functionality to combine everything into a social network. Another web is proposed in [73] called Paraimpu, which is an architecture of a large-scale social website for smart objects and services. It is a web-enabled platform which permits most of the activities between real smart objects and virtual things like web services and social networks such as adding, sharing, and connecting. A social access controller platform is covered in [70], which can act as an authentication and sharing proxy for smart things that allow users to determine what action they need for their smart things. Moreover, this platform can be used for advertising shared smart things. Additionally, web solutions to four concerns—accessibility, findability, sharing, and composition—are also presented in [65].

8.3.2.3.2 Social Cognition

The cognition of the SIoT plays a critical part in network operation and visualization. The Drive and Share (DaS) network service is presented in [74]. The DaS Machine-to-Machine platform is integrated with IP Multimedia Subsystem enablers. The DaS application supports drivers and passengers to effectively interchange information on traffic and personal data on vehicles. The research in [75] introduced a platform called the Social Devices Platform (SDP). The SDP can autonomously compose collaborated services in devices located in the same area when the user part is not complicated and is easily implementable to numerous types of devices. Moreover, a new social relations cognitive model is proposed in [76], which takes into account social changes in the relations among mobile nodes, and after that employs the information entropy and rough sets strategy to investigate the weight dissemination of social relations. The nodes social relation cognition algorithm is also of interest and is introduced in [77], which is based on the social network.

8.3.2.3.3 Location-Based Awareness
The primary direction to make SIoT accessible for mobile users is Location-Based Mobile Social Networks (LoMoSo) [74]. This idea is to share information about the location and time of mobile users around the network to enhance communication among the network's entities. An example is the DaS introduced in [74], which allows drivers and passengers, and even vehicles, to take part in the social network. The advantage of DaS is that vehicles can collect real-time road information such as traffic information, incident time, and so on, without human intervention. Hence, it allows us not only to accurately detect an incident but also to determine the duration of the incident notification. The SDP in [75] takes advantage of the physical proximity of the devices in interaction among mobile users. Similarly, [76] and [77] also use location information in their algorithm to solve shortcomings in the existing models.

8.3.2.3.4 Social Network Graph Analysis
Graph analysis has existed for a long time in analyzing any kind of network. Similarly, in SIoT, graph analysis involves a lot of promising trends. In [78], the authors used a technique to exploit graph analysis to overcome an issue in SIoT. In the method, they split the nodes in complicated networks into basic and IoT nodes. A node can be a part of many communities, and it works well for a weighted graph. A community detection algorithm is applied in the SIoT graph network.

8.3.2.3.5 Trust Management
An IoT framework interfaces a huge amount of tags, sensors, and mobile services to encourage data sharing, facilitating many appealing applications. Hence, it poses a challenge the plan and assessment of IoT systems to meet the requirements of versatility, compatibility, and dynamic flexibility. For example, a reliable system in [79] is built based on the understanding the method of providing information by the other members of the SIoT. In this research, a subjective model has been characterized for the administration of the reliability of each node's friends. Additionally, the research in [80] outlined and evaluated a reliable trust management protocol that can be expanded and adapted in dynamic IoT environments. The results have demonstrated that the trust management protocol using the constrained storage space in [80] accomplishes comparable execution with normal protocol under ideal storage conditions.

8.4 SMART CITIES AND APPLICATIONS

Thanks to recent advances in digital innovations, smart cities have become smarter. Smart cities are cities built on intelligent and knowledgeable technology, which are prepared with various electronic components, for example, sensors for traffic alert systems. The research in [81] presented some of the aspects of a smart city that have the most potential in the near future, which could be smart energy, smart building, smart mobility, smart healthcare, smart infrastructure, smart technology, smart governance, smart education, and the smart citizen. Within the IoT setting, the geographic location allows components to be integrated and evaluated by utilizing an analyzing framework. There are numerous applications that apply IoT to encourage operations in air and noise contamination tracking systems or surveillance systems. IoT affects various different aspects of the smart city citizens' life such as security, health, and transportation [82]. Moreover, it may take on a vital role in national policy decisions regarding energy saving and pollution problems. The following section summarizes technologies used in smart cities and applications and trends of IoT in smart cities.

8.4.1 IoT Technologies for Smart Cities

8.4.1.1 Radio-Frequency Identification (RFID)

RFID systems comprise readers and tags, and they play an essential part within the setting of IoT. In [83], authors integrated RFID and smart object-based architectures to form a novel IoT infrastructure. The proposed infrastructure also investigated the social characteristics of the SIoT as mentioned previously. In this structure, RFID tagged objects are considered objects that boost primitive functions, while intelligent objects are objects that support complicated functions that lead to a superset of objects. By applying these advances to any related object, it is possible to create automatic identification and assign a unique digital identity to each object within the network.

8.4.1.2 Wireless Sensor Networks (WSN)

In order to be smart, the IoT needs to deploy a number of sensors to sense the environment and send data to the center for processing and analysis. Wireless sensor networks (WSNs) can provide diverse kinds of appropriate data and can be used in numerous areas such as medical services, warning systems, and environmental services [84]. In addition, high-level objective systems are created by integrating RFID and WSNs to obtain information with respect to location, temperature, movement, and so on [82].

8.4.1.3 Addressing

Due to the very large scale of IoT smart cities, uniquely addressing the objects is a vital task to control them via the Internet. It may be considered significant for the fulfillment of the IoT infrastructure. An addressing scheme allows us to uniquely recognize billions of devices and to control remote or inaccessible devices. Every object in the network must be recognized uniquely by their identification, location, and functionalities. [85] cited that in addition to an object's unique characteristics, reliability, scalability, and persistence are the characteristics that are most necessary to create a productive addressing strategy.

8.4.1.4 Middleware

Middleware is necessary in IoT as it has the capability to precisely integrate the functionalities and communication abilities of related objects [82]. By taking advantage

Recent Trends of IoT and Big Data 155

of the IoT, a popular middleware may be accessible for the administrations or services of smart cities. In [86], a middleware is proposed to supply smart services for city administrators using cloud computing and frequency identification methods. Using this platform, cities can become smart by expanding their applications with a new set of technical, functional, and business capabilities based on the particular requirements of each city and users' demands. The framework in [87] obtained data from profoundly disseminated, heterogeneous, decentralized, real and virtual devices that can be automatically handled, examined, and controlled by cloud-based services.

8.4.2 IoT Actual Trends for Smart Cities

Thanks to the Internet, the IoT can incorporate complex devices and connect them with each other. All the involved devices are required to connect to the Internet in order to facilitate accessibility. To achieve this purpose, sensors can be executed in diverse areas for collecting and analyzing information to increase efficiency. The following briefly summarizes the main trends in this area for smart cities.

8.4.2.1 Smart Homes

In the smart home, based on an indoor IoT platform, different devices will allow common activities in the home to be operated automatically. Sensors placed in the smart home to detect such things as temperature, humidity, dust, gas, and so on will provide data about the state of the house. These sensors combine to form a sensor network, which helps smart home applications operate effectively [82]. In the study [88], people who participated in a Demand Response Program (DRP) used an energy management system to control energy consumption and reduce unnecessary energy costs. In this program, based on pollution information, people can also receive alerts about pollution if the pollution level exceeds the limit. Customers' electricity costs can be reduced by up to 48% with this model. Another study in [89] studied users' energy habits through an easy-to-use, non-invasive smart meter system to improve energy efficiency. This study solved the problem of large data volumes by building a database mechanism, electronic appliance recognition classification, and waveform recognition.

A design in [90] represents a system which can perform multi-sensing, which is applicable to real applications such as heating and air conditioning systems. In this investigation, a smart light and energy control framework is proposed using numerous sensors which can be applied in smart cities. In other studies such as [91, 92], smart lighting has been examined. Up to 45% of the energy needed for lighting could be saved by using smart lighting control mechanisms [92].

8.4.2.2 Smart Parking Lots

Parking in busy cities has been time consuming and sometimes inefficient since the number of vehicles is growing rapidly. Smart parking services are proposed to solve this problem. Smart parking systems used in [93] took advantage of a mixed-integer linear programing model for the traffic behavior of plug-in electric vehicles (PEVs). For different smart parking lots allocated within the city, arrival and departure times of various vehicles can be monitored for traffic optimization purposes. The results demonstrate the capability of the model when applied in two zones, including an urban area with residential and industrial zones. Subsequently, smart parking lots ought to be planned in such a way as to take into consideration the number of cars in each zone [94]. Besides, the parking lots are planned to allocate in distributed systems with the purpose of minimizing system costs including network reliability, power loss, and voltage deviation [95]. Smart parking lots offer not only more convenience for drivers but also advantages for businesses' daily lives.

8.4.2.3 Vehicular Traffic

In a smart city, keeping traffic moving smoothly is essential, so traffic data represents very important information. Both citizens and governments can benefit from this data [96]. Citizens can take advantage of traffic information to find the most efficient way to get to their destination. Moreover, they can even estimate the time of arrival [97] as well as be prepared for the traffic in the route. Traffic will spread evenly across the city, which reduces traffic congestion, helps reduce fuel consumption, and reduces the pollution that occurs during heavy traffic. Real-time information about traffic jams due to various causes such as accidents or other issues will help operators to take the necessary action. In [96], traffic information is obtained through the GPRS, vehicular sensors, and the sensors placed on the windshield of the car. Moreover, if any accident happens, the sensor placed in the car will send the information to the police or traffic authorities.

8.4.2.4 Environmental Pollution

Health issues for people are an essential requirement for smart city [96]. In the process of designing a smart city, a separate module for getting environmental data, for example, information on the presence of gases, ozone, sulfur dioxide, and noise is required. The module is capable of monitoring environmental pollution and providing relevant information for citizens, especially those who have health problems. Moreover, it lets governments make the right decisions at the right time regarding environmental pollution to improve citizens' quality of life.

8.4.2.5 Surveillance Systems

In the smart city, everything is connected and communicated with others, and consequently the issue of security is imperative. For this reason, everything in the city ought to be continuously monitored. However, analyzing the information and recognizing threats/crimes are challenging tasks. The research in [96] proposed new scenarios that improve security for the whole smart city. Various emergency devices in the form of buttons with attached microphones are placed in different locations in the city along with a surveillance camera system. When any violation occurs, such as robbery, car theft, fighting or any illegal activities, people can push the nearest emergency button and the data will be sent immediately to the security system.

8.4.2.6 Smart Energy and Smart Grids

We can say that the next generation of the existing power grid will be a smart grid where power distribution and management is improved by using advanced two-way communications and extensive computing capabilities to improve efficiency, safety, and reliability and guarantee real-time control [97]. The electricity in the smart gird will be delivered between suppliers and consumers by two-way technologies. This represents the modernization of the existing electricity network [98]. The term *smart energy* extends beyond the idea of smart grids to refer to the other renewable energy forms such as solar energy and wind power, in addition to electrical energy. An incredible challenge in smart energy is the integration of renewable energy subsystems into a complete decentralized system. Smart energy systems require an intelligent management system to handle the volatile state of distributed energy sources (DERs) [99]. Sensors and devices in the system rapidly continuously create data related to control and protection rings. The interaction between machines requires real-time analysis and processing to create a control request in the system. In addition, the network must also meet all visual and reporting requirements.

REFERENCES

1. N. Elgendy and A. Elragal, "Big Data analytics in support of the decision making process." *Procedia Computer Science*, 100, pp. 1071–1084, 2016.
2. H. V.. Jagadish, "Big Data and science: Myths and reality." *Big Data Research*, 2(2), pp. 49–52, 2015.
3. J. A. R. Rojas, M. B. Kery, S. Rosenthal and A. Dey, "Sampling techniques to improve Big Data exploration." In: 2017 *IEEE 7th Symposium on Large Data Analysis and Visualization (LDAV)*, Phoenix, AZ, USA, 2017.
4. M. Offroy and L. Duponchel, "Topological data analysis: A promising Big Data exploration tool in biology, analytical chemistry and physical chemistry." *Analytica Chimica Acta*, 910(3), pp. 1–11, 2016.
5. S. Li, S. Dragicevic, F. Anton and M. Sester, "Geospatial Big Data handling theory and methods: A review and research challenges." *ISPRS Journal of Photogrammetry and Remote Sensing*, 2015.
6. N. Bikakis and T. Sellis, "Exploration and visualization in the web of big linked data: A survey of the state of the art." In: *6th International Workshop on Linked Web Data Management (LWDM 2016)*.
7. S. StrauB, "Datafication and the seductive power of uncertainty —A critical exploration of Big Data enthusiasm." *Information*, 6(4), pp. 836–847, 2015.
8. D. P. Acharjya and K. Ahmed, "A survey on Big Data analytics: Challenges, open research issues and tools." *(IJACSA) International Journal of Advanced Computer Science and Applications*, 7(2), pp. 511–518, 2016.
9. A. Bifet, G. Holmes, R. Kirkby and B. Pfahringer, "MOA: massive online analysis." *The Journal of Machine Learning Research: JMLR*, 11, pp. 1601–1604, 2010.
10. Amazon, "AWS case study: SwiftKey." 2014 [Online] Available: http://aws.amazon.Com/Solutions/Case-Studies/Big-Data/. [Accessed 3, 8, 2014].
11. Cloudera, "Using Big Data to bridge the virtual & physical worlds." 2014 [Online]. Available: http://www.cloudera.com/content/dam/cloudera/documents/Cloudera-Nokia-case-study-final.pdf. [Accessed 23, 7, 2014].
12. Alacer, "Case studies: Big Data." 2014 [Online]. Available: http://www.alacergroup.Com/practice-category/big-data/casestudies/. [Accessed 24, 7, 2014].
13. I. Yaqoob, I. A. T. Hashem, A. Gani, S. Mokhtar, A. Ahmed, N. B. Anuar and A. V. Vasilakos, "Big Data: From beginning to future." *International Journal of Information Management*, 36(6), pp. 1231–1247, 2016.

14. H. Li, G. Fox and J. Qiu, 2012, "Performance model for parallel matrix multiplication with dryad: Dataflow graph runtime." In: *Proceedings of the 2012 Second International Conference on Cloud and Green Computing*, pp. 675–683.
15. C. W. Tsai, "Big Data analytics: A survey." *Journal of Big Data*, 2(1), pp. 1–32, 2015.
16. V. Estivill-Castro, "Why so many clustering algorithms: A position paper." *ACM SIGKDD Explorations Newsletter*, 4(1), pp. 65–75, 2002.
17. C. Bielza and P. Larrañaga, "Discrete Bayesian network classifiers: A survey." *ACM Computing Surveys*, 47(1), 2014.
18. R. A. A. D. Luss, "Predicting abnormal returns from news using text classification." *Quantitative Finance*, 15(6), pp. 999–1012, 2015.
19. P. A. O. C. Melin, "A review on type-2 fuzzy logic applications in clustering, classification and pattern recognition." *Applied Software Computing*, pp. 568–577, 2014.
20. A. Soualhi, K. Medjaher and A. N. Zerhouni, "Bearing health monitoring based on Hilbert–Huang transform, support vector machine, and regression." *IEEE Transactions on Instrumentation and Measurement*, 64(1), pp. 52–62, 2015.
21. C. C. Aggarwal, *Recommender Systems: The Textbook*, Springer; 1st ed, 2016.
22. V. Borkar, M. Carey and A. A. Li, "Inside "Big Data management": Ogres, onions, or parfaits?" In: *Proceedings of the 15th International Conference on Extending Database Technology, EDBT*, pp. 3–14, 2012.
23. A. E. A. Gani, "A survey on indexing techniques for Big Data: Taxonomy and performance evaluation." *Knowledge and Information Systems*, pp. 241–284, 2016.
24. A. E. A. Paul, "Video search and indexing with reinforcement agent for interactive multimedia services." *ACM Transaction on Embedded Computer Systems*, 12(2), pp. 1–16, 2013.
25. P. Berkhin, "A survey of clustering data mining techniques." In: *Grouping Multidimensional Data*, Springer, pp. 25–71, 2006.
26. A. Fahad, N. Alshatri, Z. Tari, A. Alamri, I. Khalil, A. Y. Zomaya, S. Foufou and A. Bouras, "A survey of clustering algorithms for Big Data: Taxonomy and empirical analysis." *IEEE Transactions on Emerging Topics in Computing*, 2(3), pp. 267–279, 2015.
27. A. Gosain and M. Bhugra, "A comprehensive survey of association rules on quantitative data in data mining." In: *IEEE Conference on Information & Communication Technologies*, Thuckalay, Tamil Nadu, India, 2013.
28. L. C. A. M. C. Soon, "Efficient mining of maximal sequential patterns using multiple samples." In: *Proceedings of the 2005 SIAM International Conference on Data Mining*, 2005.
29. Z. A. M. K. Yang, "LAPIN-SPAM: An improved algorithm for mining sequential pattern." In: *21st International Conference on Data Engineering Workshops (ICDEW'05)*, 2005.
30. A. Gandomi and M. Haider, "Beyond the hype: Big Data concepts, methods, and analytics." *International Journal of Information Management*, 35(2), pp. 137–144, 2015.
31. K. Kalpakis, D. Gada, V. Puttagunta. "Distance measures for effective clustering of ARIMA time- series." In: *Proceedings of the IEEE International Conference on Data Mining*, 2001.
32. N. Kumar. "Time-series bitmaps: A practical visualization tool for working with large time series databases." In: *SDM*, 2005.
33. D. Ryan. *High Performance Discovery in Time Series: Techniques and Case Studies*, Springer Science & Business Media., 2013.
34. R. R.-C. U. O. Malaga, J. López and S. Gritzalis, "Evolution and trends in IoT security." *Computer*, 51(7), 2018.
35. J. H. Kim, "A survey of IoT security: Risks, requirements, trends, and key technologies." *The Journal of Industrial Integration and Management*, 2(2), 2017.

36. M. Akgun and M. U. Caglayan, "Providing destructive privacy and scalability in RFID systems using PUFs." *Science Direct*, 32, pp. 32–42, 2015.
37. H.-Y. Chien, T.-C. Wu and C.-L. HSU, "RFID authentication with un-traceability and forward secrecy in the partial-distributed-server model." *IEICE Transactions on Information and Systems*, E98.D(4), pp. 750–759, 2015.
38. B. R. Ray, J. Abawajy and M. Chowdhury, "Scalable RFID security framework and protocol supporting Internet of Things." *Computer Networks*, 67, pp. 89–103, 2014.
39. N. Park and N. Kang, "Mutual authentication scheme in secure Internet of Things technology for comfortable lifestyle." *Sensors*, 16(1), 2016.
40. K. J. G. Park and, J. Park, "Design of secure authentication scheme between devices based on zero-knowledge proofs in home automation service environments." *The Journal of Supercomputing*, 72(11), pp. 4319–4336, 2016.
41. T. Shinzaki, I. Morikawa, Y. Yamaoka and Y. Sakemi, "IoT security for utilization of Big Data: Mutual authentication technology and anonymization technology for positional data." *Fujitsu Scientific and Technical Journal*, 52(4), pp. 52–60, 2016.
42. Y. Hong, J. Kim and D. Kim, "A proposal of multi-factor authentication scheme for secure IoT environment." *ICIC Express Letters Part B, Applications: An International Journal of Research and Surveys*, 6(12), pp. 3231–3236, 2015.
43. G. Xu, Y. Han, Y. Ren and, X. Li, "Privacy protection method based on two-factor authentication protocol in FRID systems." *IEICE Transactions on Information and Systems*, pp. 2019–2026, 2016.
44. M. Almulhim and N. Zaman, "Proposing secure and lightweight authentication scheme for IoT based e-health applications." In: *20th International Conference on Advanced Communication Technology (ICACT)*, Chuncheon-si Gangwon-do, Korea, 2018.
45. J.-h. Han and J. Kim, "A lightweight authentication mechanism between IoT devices." In: *International Conference on Information and Communication Technology Convergence (ICTC)*, Jeju, South Korea, 2017.
46. K. Fan, Y. Gong, C. Liang, H. Li and Y. Yang, "Lightweight and ultralightweight RFID mutual authentication protocol with cache in the reader for IoT in 5G." *Wiley Security and Communication Networks*, 2015.
47. S. Li, T. Tryfonas and H. Li, "The Internet of Things: A security point of view." *Internet Research*, 26(2), pp. 337–359, 2016.
48. S. Peng and H. Shen, "Security technology analysis of IoT." *Internet of Things, Springer*, pp. 401–408, 2012.
49. J. Bokefode, A. Bhise, P. Satarkar and D. Modani, "Developing a secure cloud storage system for storing IoT data by applying role based encryption." *Procedia Computer Science*, 89, pp. 43–50, 2016.
50. B. V. Sundaram, M. Ramnath, M. Prasanth and J. V. Sundaram, "Encryption and hash based security in Internet of Things." In: *3rd International Conference on Signal Processing, Communication and Networking (ICSCN)*, Chennai, India, 2015.
51. M. Elhoseny, G. Ramírez-González, O. M. Abu-Elnasr, S. A. Shawkat, A. N. and A. Farouk, "Secure medical data transmission model for IoT-based healthcare systems." *IEEE Access*, 6, pp. 20596–20608, 2018.
52. M. Elhoseny, K. Shankar, S. K. Lakshmanaprabu, A. Maseleno and N. Arunkumar, "Hybrid optimization with cryptography encryption for medical image security in Internet of Things." *Springer Computer Aided Medical Diagnosis*, 2019.
53. M. Khari, A. K. Garg, A. H. Gandomi, R. Gupta, R. Patan and B. Balusamy, "Securing data in Internet of Things (IoT) using cryptography and steganography techniques." In: *IEEE Transactions on Systems, Man., and Cybernetics: Systems (Early Access)*, pp. 1–8, 2019.
54. M. Grabovica, S. Popić, D. Pezer and V. Knežević, "Provided security measures of enabling technologies in Internet of Things (IoT): A survey." In: *Zooming Innovation in Consumer Electronics International Conference*, Novi Sad, Serbia, 2016.

55. H. J. Ban, J. Choi and N. Kang, "Fine-grained support of security services for resource constrained Internet of Things." *International Journal of Distributed Sensor Networks*, 2016.
56. M. Sethi, J. Arkko and A. Keranen, "End-to-end security for sleepy smart object networks." In: *IEEE 27th Conference on Local Computer Networks Workshops (LCN Workshops)*.
57. K. H. Han and W. Bae, "Proposing and verifying a security-enhanced protocol for IoT-based communication for medical devices." *Cluster Computing*, 19(4), pp. 2335–2341, 2016.
58. A. G. Steri, G. Baldini, I. N. I. Fovino, R. Neisse and L. Goratti, "A novel multi-hop secure LTE-D2D communication protocol for IoT scenarios." In: *23rd International Conference on Telecommunications (ICT)*, Thessaloniki, Greece, 2016.
59. Z.-K. Zhang, M. C. Y. Cho, C.-W. Wang, C.-W. Hsu, C.-K. Chen and S. Shieh, "IoT security: Ongoing challenges and research opportunities." In: *IEEE 7th International Conference on Service-Oriented Computing and Applications*, Matsue, Japan, 2014 .
60. J. Zheng and D. SimPlot-Ryl, C. Bisdikian and A. H. Mouftah, "The Internet of Things." *IEEE Communications Magazine*, 49(11), pp. 30–31, 2011.
61. A. M. Ortiz, D. Hussein, S. Park, S. N. Han and N. Crespi, "The cluster Between Internet of Things and social networks: Review and research challenges." *IEEE Internet of Things Journal*, 1(3), pp. 206–215, 2014.
62. M. Kranz, F. Michahelles and L. Roalter, "Things that Twitter: Social networks and the Internet of Things." In: *What Can the Internet of Things Do for the Citizen (CIoT) Workshop at the Eighth International Conference on Pervasive Computing*, 2019.
63. M. Baqer and A. Kamal, "S-sensors: Integrating physical world inputs with social networks using wireless sensor networks." In: *Proceedings of the 5th International Conference Intell. Sens. Sens. Netw. Inf. Process. (ISSNIP)*, pp. 213–218, 2009.
64. M. Baqer, "Enabling collaboration and coordination of wireless sensor networks via social networks." In: *6th IEEE International Conference on Distributed Computing in Sensor Systems Workshops (DCOSSW)*, Santa Barbara, CA, USA, 2010.
65. D. Guinard, "A Web of Things Application Architecture: Integrating the Real World into the Web." Ph.D dissertation, Zurich, Switzerland, 2011.
66. L. Atzori, A. Iera, G. Morabito and M. Nitti, "The Social Internet of Things (SIoT)—When social networks meet the Internet of Things: Concept, architecture and network characterization." *Computer Networks*, 56(16), pp. 3594–3608, 2012.
67. L. Atzori, D. Carboni and A. Iera, "Smart things in the social loop: Paradigms, technologies, and potentials." *Ad Hoc Networks*, 18, pp. 121–132, 2014.
68. L. Atzori, A. Iera and G. Morabito, "SIoT: Giving a social structure to the Internet of Things." *IEEE Communication Letter*, 15(11), pp. 1193–1195, 2011.
69. J. I. Vazquez and D. Lopez-de-Ipina, "Social devices: Autonomous artifacts that communicate on the Internet." *The Internet of Things*, pp. 308–324, 2008.
70. D. Guinard, V. Trifa and M. Fischer, 2010, "Sharing using social networks in a composable Web of Things." In: *8th IEEE International Conference on Pervasive Computing and Communications Workshops (PERCOM Workshops)*.
71. A. Ciortea, O. Boissier, A. Zimmermann and A. M. Florea, "Reconsidering the social web of things. Position Paper." In: *Proceedings of the 2013 ACM conference on Pervasive and Ubiquitous Computing Adjunct Publication*, Zurich, Switzerland, 2013.
72. H. Ning and Z. Wang, "Future Internet of Things architecture: Like mankind neural system or social organization framework?" *IEEE Communications Letters*, 15(4), pp. 461–463, 2011.
73. A. Pintus, A. Piras and D. Carboni, "Paraimpu: A platform for a social Web of Things." In: *Proceedings of the 21st Annual Conference on World Wide Web Companion*, 2012.

74. I. Lequerica, M. G. Longaron and P. M. Ruiz, "Drive and share: Efficient provisioning of social networks in vehicular scenarios." *IEEE Communications Magazine*, 48(11), pp. 90–97, 2010.
75. N. Mäkitalo, J. Pääkkö, M. Raatikainen, V. Myllärniemi, T. Aaltonen, T. Leppänen, T. Männistö and T. Mikkonen, "Social devices: Collaborative co-located interactions in a mobile cloud." In: *Proceedings of the 11th International Conference on Mobile and Ubiquitous Multimedia*, Ulm, Germany, 2012.
76. J. An, X. Gui, W. Zhang, J. Jiang and J. Yang, "Research on social relations cognitive model of mobile nodes in Internet of Things." *Journal of Network and Computer Applications*, 36(2), pp. 799–810, 2013.
77. J. An, X. Gui, W. Zhang and J. Jiang, "Nodes social relations cognition for mobility-aware in the Internet of Things." In: *International Conference on Internet of Things and 4th International Conference on Cyber, Physical and Social Computing*, Dalian, China, 2011.
78. S. Misra, R. Barthwal and M. S. Obaidat, "Community detection in an integrated Internet of Things and social network architecture." In: *IEEE Global Communications Conference (GLOBECOM)*, Anaheim, CA, USA, 2012.
79. I. Michele Nitti, University of Cagliari, R. Girau, L. Atzori, A. Iera and G. Morabito, "A subjective model for trustworthiness evaluation in the social Internet of Things." In: *IEEE 23rd International Symposium on Personal, Indoor and Mobile Radio Communications—(PIMRC)*, Sydney, NSW, Australia, 2012.
80. F. Bao, I.-R. Chen and J. Guo, "Scalable, adaptive and survivable trust management for community of interest based Internet of Things systems." In: *IEEE Eleventh International Symposium on Autonomous Decentralized Systems (ISADS)*, Mexico City, Mexico, Mexico, 2013.
81. "Strategic opportunity analysis of the global smart City Market." [Online]. Available: https://dsimg.ubm-us.net/envelope/153353/295862/1391029790_strategic_opportunity.pdf.
82. H. Arasteh, V. Hosseinnezhad, V. Loia, A. Tommasetti, O. Troisi, M. Shafie-khah and P. Siano, "IoT-based smart cities: A survey." In: *IEEE 16th International Conference on Environment and Electrical Engineering (EEEIC)*, Florence, Italy, 2016.
83. A. K. Evangelos, D. T. Nikolaos, C. B. Anthony, "integrating RFIDs and smart objects into a unified Internet of Things architecture." *Advances in Internet of Things*, 1, pp. 5–12, 2011.
84. W. S. Ansari, A. M. Alamri, M. M. Hassan and M. Shoaib, "A survey on sensor-cloud: Architecture, applications, and approaches." *International Journal of Distributed Sensor Networks*, 2013.
85. J. Gubbi, R. Buyya, S. Marusic and M. Palaniswami, "Internet of Things (IoT): A vision, architectural elements, and future directions." *Future Generation Computer Systems*, 29(7), pp. 1645–1660, 2013.
86. P. Ballon, P. Kranas, J. Glidden, A. Menychtas, S. Ruston and S. v. d. Graaf, "Is there a need for a cloud platform for European smart cities?" In: *eChallenges e-2011 Conference Proceedings, IIMC International Information Management Corporation*, pp. 1–7, 2011.
87. G. Suciu, A. Vulpe, S. Halunga, O. Fratu, G. Todoran and V. Suciu, "Smart cities built on resilient cloud computing and secure Internet of Things." In: *19th International Conference on Control Systems and Computer Science*, Bucharest, Romania, 2013.
88. M. Shafie-khah, M. Kheradmand, S. Javadi, M. Azenha, J. d. Aguiar, J. Castro-Gomes, P. Siano and J. Catalão, "Optimal behavior of responsive residential demand considering hybrid phase change materials." *Applied Energy*, 163(1), pp. 81–92, 2016.
89. S.-Y. Chen, C.-F. Lai, Y.-M. Huang and Y. L. Jeng, "Intelligent home-appliance recognition over IoT cloud network." In: *9th International Wireless Communications and Mobile Computing Conference (IWCMC)*, Sardinia, Italy, 2013.

90. D.-M. Han and J.-h. Lim, "Smart home energy management system using IEEE 802.15.4 and zigbee." *IEEE Transactions on Consumer Electronics*, 56(3), pp. 1403–1410, 2010.
91. X. Ye and J. Huang, "A framework for Cloud-based smart home." In: *International Conference on Computer Science and Network Technology (ICCSNT)*, vol. 2, pp. 894–897, 2011.
92. L. Martirano, "A smart lighting control to save energy." In: *IEEE 6th International Conference on Intelligent Data Acquisition and Advanced Computing Systems (IDAACS)*, vol. 1, pp. 132–138, 2011.
93. N. Neyestani, M. Shafie-khah, M. Y. Damavandi and J. P. S. Catalão, "Modeling the PEV traffic pattern in an urban environment with parking lots and charging stations." In: *IEEE Power Technical Conference*, Eindhoven, Netherlands, 2015.
94. M. Yazdani-Damavandi, M. P. Moghaddam, M.-R. Haghifam, M. Shafie-khah and J. P. S. Catalão, "Modeling operational behavior of plug-in electric vehicles' parking lot in multienergy systems." *IEEE Transactions on Smart Grid*, 7(1), pp. 124–135, 2016.
95. N. Neyestani, M. Y. Damavandi, M. Shafie-Khah, J. Contreras and J. P. S. Catalão, "Allocation of plug-in vehicles' parking lots in distribution systems considering network-constrained objectives." *IEEE Transactions on Power Systems*, 30(5), pp. 2643–2656, 2015.
96. M. M. Rathore, A. Ahmad, A. Paul and S. Rho, "Urban planning and building smart cities based on the Internet of Things using Big Data analytics." *Computer Networks*, 101, pp. 63–80, 2016.
97. M. Shafie-khah, E. Heydarian-Forushani, M. Golshan, P. Siano, M. Moghaddam, M. Sheikh-El-Eslami and A. J. Catalão, "Optimal trading of plug-in electric vehicle aggregation agents in a market environment for sustainability." *Applied Energy*, 162, pp. 601–612, 2016.
98. Y. Yan, Y. Qian, H. Sharif and D. Tipper, "A survey on smart grid communication infrastructures: Motivations, requirements and challenges." *IEEE Communications Surveys and Tutorials*, 15(1), pp. 5–20, 2013.
99. E. Nada and E. Ahmed, "Big Data analytics: A literature review paper." *Advances in Data Mining, Applications and Theoretical Aspects*, pp. 214–227, 2014.

9 A Theoretical Context for CSF in Medical Software Next Release

Hemlata Sharma, Shushila Madan, Nisheeth Joshi, and Nidhi Mathur

CONTENTS

9.1 Introduction 163
9.2 Background 164
    9.2.1 Software NRP 164
    9.2.2 Critical Success Factor Approach 165
9.3 Success Factors for NRP-Based Software Projects 166
9.4 Research Methodology 167
9.5 Data Analysis and Results 167
    9.5.1 Exploratory Factor Analysis 169
        9.5.1.1 Multiple Regression Model 170
    9.5.2 Discussion 173
9.6 Conclusions 174
9.7 Limitations and Future Scope 178
Appendix 180
References 181

9.1 INTRODUCTION

The software industry is undergoing several fundamental shifts due to current economic necessities and ongoing technological developments. Software companies need to develop software products that are reliable and cost effective and can be delivered on time so that they can meet the demands of their customers and stakeholders efficiently. As per Johnson's report [1], a project may lead to unsatisfactory results if it is badly researched or developed. One possibility could be the incorrect identification of needs/features/requirements, which may lead to inaccurate results. Many past studies [2, 3] have shown that if a software project is inaccurately developed, it leads to dissatisfaction on the part of end users. Therefore, the requirements should be identified and employed correctly so that end users' satisfaction and benefit would be maximized. However, the identification of the exact requirement specifications of end users is a difficult task, due to the exponential growth of requirements. Thus, high development and maintenance costs are involved in determining an appropriate set of requirements given by various stakeholders. At times, it has been noticed that some requirements are not met or accomplished for one reason or another. Therefore, developers aim to upgrade the software by the addition of these requirements in the next release, which are specifically designed to fulfill the requirements and fix the problems identified in the current release for next or future releases. The *next release problem* (NRP) illustrates this issue by pointing out those software requirements which should be incorporated in the next release of any software product. Bagnall et al. [4] called this problem the software next release problem. The aim of this is to accurately adjust plans and cost estimates to obtain maximum revenue. NRP is a complex optimization problem [5]. Techniques like requirement dependencies and requirement prioritization have been applied by various researchers [6, 7] for requirement selection while solving NRP. The NRP is framed as single-objective [4], and multi-objective [8] problem, in order to fulfill the company's required objective.

The current research is trying to find out the critical success factors (CSF) required for the success of software NRP. The study will provide meaningful insights for these CSF. This study proposes a conceptual framework for the key factors influencing the success of projects based on the NRP approach. Web-based and individual surveys were conducted to collect responses from 266 software professionals involved in NRP. The study evaluated the success factors described in the literature for NRP; factor and reliability analysis was then done on these extracted sets of factors, and the factors were presented in the form of seventeen potential CSF for NRP-based software projects in five different groups: people, organizational, planning, implementation, and testing. This chapter presents the background to the problem followed by the research methodology. The framework is then proposed based on the identified critical success factors. Concluding remarks are followed by a discussion of the future scope and limitations of the study.

9.2 BACKGROUND

NRP faces many challenges derived from difficulties like communication and cultural differences which may lead to a software disaster. It is evident from the literature that if requirements are not satisfactorily met, companies may face problems like losing important customers, over-budgeting, and failing to release software projects on time. The scope of NRP planning is limited because of how fast requirements change. Hence, it is important to establish some factors by which the success of NRP-based software could be identified. This section highlights the background to these two key concepts in brief: NRP-based software projects and CSF.

9.2.1 SOFTWARE NRP

The *next release* of a software product is a group of new or/and modified requirements. Iterative and incremental software development is in high demand as designers come up with various additive functionalities known as *releases*. These releases are also known as *versions*. The NRP is a problem of deciding the various features to be added in the next release of a software product to satisfy stakeholders' requirements. Each stakeholder may have a requirement for a particular feature. Thus, release management practices must be as good as possible, so we can guarantee that when we build a software product, it will be efficaciously distributed to the stakeholders or clients who want to make use of it and that these requirements are optimized so that maximum revenue can be generated. In so doing, the software company can not only satisfy existing clients but also attract new ones. However, requirement selection is a difficult and complex task as it is tough to identify the valued user, the market impact, the dependencies between requirements, and so on.

Adams *et al.* [9] interviewed three release engineers to find out the real-time practices and challenges faced by release engineers. 80% of the firms that subcontract their growth and maintenance had issues. 20% of sourcing agreements were cancelled within one year. 50% contracts could not achieve their aim. Hence, these projects were terminated [10]. According to Ebert *et al.* [11], a product's success criteria are those features which differentiate it from other products, rather than the number of features it has. The major problem with NRP is uncertainty [12], which could be significantly affected by proposed feasible solutions which in turn could drive risk while planning. Bagnall *et al.* [4] pointed out software NRP as a major challenge for a company because of its complex selection of specific features. Therefore, identification of success factors in software NRP would certainly maximize the company's revenue. Software NRP not only helps to reduce the budget and time to deliver the project to market, but also ensures the improvement of the reliability, maintainability, and quality of the software product. The software development life cycle (SDLC) and software engineering iterative model of software development is different from that of the traditional ones. An iterative model [13, 14] shows six steps for successful iterative development.

Figure 9.1 depicts the steps of the iterative model discussed by Aggarwal and Singh [15]. The authors in their work highlighted that these six steps play a vital role during the next release/next iteration of a software implementation. These steps cannot be ignored during the whole process of the software development life cycle and can increase the chances of implementing and distributing successful software based on NRP optimization. Steps for an iterative enhancement model are depicted in Figure 9.1.

This study includes all the six stages of the iterative model for an NRP software solution to achieve the CSFs that lead to its success.

9.2.2 Critical Success Factor Approach

The literature suggests that 80% of information technology (IT) projects go over budget, 90% fail to achieve their objective, and 40% are discarded [16]. The Standish group proposed that 31% of projects are cancelled before completion and 52.7% of the projects completed go over budget by 189% [17]. IT is the backbone of today's era. It is highly desirable and important that the implementations of IT projects are done with the utmost care. Research suggests that there are many factors which are important for the success of IT projects. These factors are known as critical success

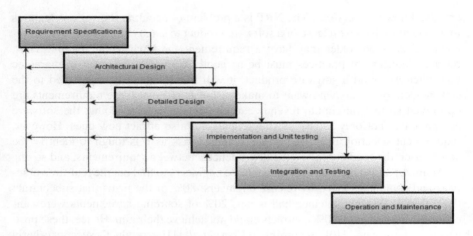

FIGURE 9.1 Six steps for the iterative enhancement model.

factors (CSF). Determining CSF for software development projects is of great importance in today's scenario. Managers, developers, and any other authority can use them for fulfilling their goals and to achieve the desired result, that is, software which is acceptable to the customer and is developed by utilizing optimal resources, hence increasing the organization's bottom line.

Prominent studies have highlighted that the concept of CSF is very important in software development. According to Reel [17], CSF for a software project are associated with project management procedures, whereas other researchers [18] identify CSF as a combination of business strategy and software engineering. Further in this direction, Bosghossian et al. [19] suggested that the CSF of software projects is a combination of different factors: the software development life cycle, validation, estimation, resource management, executive management, and project management. Chow and Cao [20] used a quantitative approach to determining the CSF in agile software development. An extensive literature survey [21] indicated the CSF of a software project comprises a comprehensive analysis of requirements and precise estimation of allotted budget and schedule, along with a proficient project manager. CSF is gaining popularity because of its versatile fields of study. Contemporary research has used CSFs in disaster management [22], mobile learning platforms [23], and front-end development [24], to name a few.

9.3 SUCCESS FACTORS FOR NRP-BASED SOFTWARE PROJECTS

No prominent study has been encountered in the literature on CSF for NRP-based software projects. However, there are lots of papers which talk about the important activities that must take place during software development using an NRP-based approach. Some of them are described here. Van den Akker et al. [25] identified team capability, available resources, period, and maximum revenue. Development time and desirability to the customer was discussed by Bagnall et al. [3].

Researchers worked on requirement dependencies ([26] and [27]). Ricard *et al.* [28] identified technical abilities for proficiency levels related to the profile of software professionals. Working on NRP-based software development [29] resources, along with the importance of the requirements from the customer's perspective, were also considered. Uncertainties and stakeholders' valuations were highlighted as important criteria [12]. Veerapen *et al.* [30] pointed out cost and profit. Based on the discussed literature, NRP-based software projects can be categorized into five classes: people, organizational, planning, implementation, and testing, as summarized in Table 9.1.

The parameters of success related to software next release projects taken from the literature are shown in Table 9.2. These attributes affect the perceived success of NRP.

9.4 RESEARCH METHODOLOGY

The data was collected through questionnaires distributed to software professionals who are working on NRP-based software projects. Responses to a web survey with a Likert scale (from 1 = Strongly Disapprove to 7 = Strongly Approve) with questions interrelated to NRP software development and demographic information of the respondents were collected from the sample respondents from the target population. There were three segments in the questionnaire. The first segment aimed to collect demographic information regarding the NRP-based software project. The second segment was designed to find out the importance of success factors on a seven-point Likert scale. The third section measured the sensitivity of the success of the NRP-based software project; a seven-point Likert scale was used to reflect the opinion of the respondents.

To avoid vagueness in terms of parameters for success, the respondents were asked to focus only on the latest project which he/she had completed successfully.

A preliminary study to examine the content reliability and validity of the questionnaire was sent to 40 respondents, and, based on their feedback, the questionnaire was further improved before the survey invitation was sent to the target respondents. After seven weeks of survey time, 159 people had replied through reading the online survey and 107 respondents had filled in the complete form. Table 9.3 shows the role, experience, and team size of the respondents respectively.

9.5 DATA ANALYSIS AND RESULTS

A reliability analysis is done to determine the extent of reliability in all the factors. Reliability analysis delivers a general index of internal stability of the variables. Rubin and Babbie [31] found the coefficient alpha method to be a powerful method for computing the reliability and internal stability of the variables. To determine this, it was decided that a coefficient alpha level of 0.5 is acceptable [32]. Cronbach's alpha method was used to perform reliability analysis on all the items. The result of Cronbach's α test was 0.725, which is considered to be good. The closer the value of α to 1, the better the reliability test [33]. This section is further categorized into two parts: one is exploratory factor analysis, and the other is multiple regression models.

TABLE 9.1
List of Success Factors Identified from the Literature

Dimension	Factor
People	1. Team members have high levels of competence
	2. Smaller teams deliver high quality software
	3. Inter-team coordination
	4. Team members have coherent self-organizing behavior
	5. Plans are communicated to team members who worked on previous iterations
	6. Team members must have worked on previous iteration
	7. Known working speed of team is important
	8. Committed manager is responsible for successful completion
	9. Sensitive towards customer's expectations
	10. Organization is sensitive towards customer retention
	11. Team members must be free to work flextime
	12. Team members are flexible in their work
	13. Customer's regular involvement is essential during software development process
Organizational	14. Clear goals and objectives
	15. Well defined organizational structure
	16. Organization works on principle of producing NRP which will be open to work with other systems.
	17. Follows Software Engineering Standards
	18. Cooperative structure is better than hierarchical
	19. Organizational politics impact on performance
	20. Innovation and creativity
	21. Employee reward system
	22. Organization believes in key role of manager as a team motivator
Planning	23. Evaluate requirements on the basis of post-release maintenance
	24. Organization's focus is ready with alternate plan
	25. Organization's focus is on regularization of results
	26. Try to satisfy maximum requirements given by various stakeholders within given budget
	27. Client's technical skills helps in clarity while planning
	28. Backlogs should be pre-assigned to the team
	29. Writing of requirement was done beforehand
	30. Plans to ensure maximum resource utilization
	31. Release deadlines are normally fixed
	32. Release software development cycle should be kept as short as possible
	33. In case of long release, better to break it into parts
	34. Fast decision in release planning is actually important
	35. Evaluate requirements based on execution time
	36. Evaluate requirements on the basis of dependency between requirements
	37. Evaluate requirements on the basis of schedule estimates
	38. Evaluate requirements on the basis of priority
	39. Evaluate requirements on the basis of cost estimates
	40. For selecting a particular requirement, is product's delivery time considered to be an important criterion

(continued)

CSF for Software Next Release

TABLE 9.1 (CONTINUED)
List of Success Factors Identified from the Literature

Dimension	Factor
	41. Current version is better than previous version
	42. Advance planning is essential for monitoring and risk-management plans
	43. Evaluate requirements on the basis of project development approach (build/buy)
	44. Market Analyst Perspective (valuation as per market scenario) about the requirement is an important criterion for requirement selection
	45. Leaving no slack time creates pressure on team
Implementation	46. Corrective measures were taken due to change in technology
	47. Corrective measures were take due to change in stakeholders' requirements
	48. Corrective measures were taken place due to change in market demand
	49. NRP must be able to change as per business requirements
	50. Organization is maintaining proper documentation
	51. NRP would give high productivity, hence high profit
	52. Requirement should be fully understood and defined
	53. Appropriate technical training has been given to accomplish a task
	54. Significant quality management plans are considered for requirement development
Testing	55. Must be robust
	56. Feature compatibility issues are handled
	57. Must be close to user requirements
	58. Unit and integration testing has been done correctly
	59. Must be user friendly
	60. Must be adaptable
	61. Functional security testing
	62. Security on white box analysis
	63. Company provides efforts to reduce fraud
	64. Security check by passing malicious code
	65. Must be reliable

9.5.1 Exploratory Factor Analysis

A prime component factor analysis was accomplished on the data items with varimax rotation. Seventeen factors emerged from the analysis, accounting for 76.317% of the variance. Items with loadings ≥ 0.5 on the target construct were retained. The outcome of this analysis is shown in Table 9.7, which contains information on reliability coefficients and factor loadings for each scale. Sixty-eight main hypotheses are generated from the 17 factors, each associating its presence as a CSF in terms of four perceived success dimensions: time, cost, quality, and scope.

The proposed framework is shown in Figure 9.2, which diagrammatically represents the final list of 17 factors.

There are 17 hypotheses numbered as 1, 2, 3, ... 17, and as we discussed earlier about the four perceived success dimensions for each factor, that is why there are 68 total hypotheses, as 17×4=68. These four perceived success dimensions for each

TABLE 9.2
Success Dimension

Success Dimension	Success Parameters
Perceived success parameters	1. Time (on-time delivery) 2. Cost (cost-effective delivery of product) 3. Quality (delivered a quality product) 4. Scope (all requirements and objectives should be fulfilled)

TABLE 9.3
Respondents' Demographic Information

Role of Respondent in the Organization	Number	Percentage
Project managers	84	31.5%
Software developers	93	34.9%
Test engineers	82	30.8%
Project Experience (Years)	**Number**	**Percentage**
< 3	86	32.3%
3–5	91	34.2%
6–8	78	29.3%
> 8	11	4.1%
Team Size	**Number**	**Percentage**
< 8	108	40.6%
8–15	63	23.6%
16–20	39	14.6%
21–25	56	21.0%

individual factor were indicated by the letters a, b, c, and d. The list of 68 hypotheses can be seen in Appendix from 1a to 17d.

9.5.1.1 Multiple Regression Model

Multiple regression analysis is used to determine those factors which could have positive impact on the success of a software project based on NRP optimization. It basically explores the relationship between multiple independent variables (success factors) and the dependent variable (NRP software project success). The multiple regression model with k independent variables is given in Equation 9.1.

$$Z = \beta_0 + \beta_1 x_1 + \beta_2 x_2 + \ldots + \beta_n x_n + \varepsilon \qquad (9.1)$$

where
  ε: Random error component
  β_i: Regression coefficient
  $x_1, x_2 \ldots x_n$: Independent variables
  Z: Dependent variable

CSF for Software Next Release

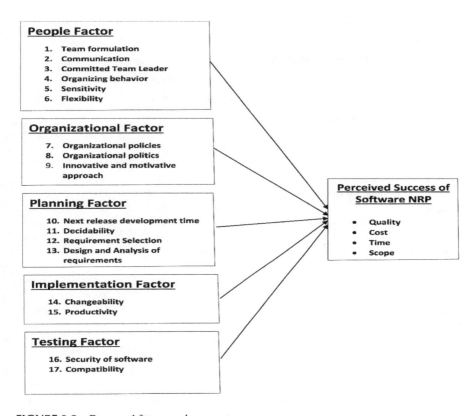

FIGURE 9.2 Proposed framework.

The value of the coefficient β_i determines the contribution of independent variable x_i, given that β_0 is the y intercept and other x variables are constant.

The research interprets the above work given by the following general equation:

$$Z(T,C,Q,S) = \beta_1 QF_1 + \beta_2 QF_2 + \ldots + \beta_{17} QF_{17} \tag{9.2}$$

In the same way as in Equation 9.2, the other three dimensions (scope, time, and cost) are calculated simply by replacing Q with S, T, and C. Where Z is the dependent variable that denotes the success of an NRP-based software project, the quality dimension is represented by Q and the scope dimension is defined by S. However, the time and cost dimensions are defined by T and C, respectively. The partial regression coefficient is shown by β_i for the ith success factor (Table 9.4).

A step-wise regression was performed by taking all 17 factor scores as the independent variables and the dependent variable as one of the four success dimensions mentioned above. Variables having a significant impact on the dependent variable would be considered as the CSF.

The results show that for the success dimension of quality, out of 17 success factors, 12 factors supported quality as a success dimension: design and analysis of requirements, requirement selection changeability, organizational policies, next-release development

TABLE 9.4
Results of the Regression Analysis

	Dependent Variable			
Independent Variable	Quality	Scope	Time	Cost
Design and analysis of requirements	.431**	.308**	.218**	.416**
Security of software				.133**
Requirement selection	.167**	.141**	.197**	.354**
Changeability	.314**	.312**	.137**	.192**
Organizational policies	.135**	.288**	.287**	
Next-release development time	.179**	.184**	.159**	.212**
Compatibility	.232**	.248**	.267**	.219**
Innovative and motivative approach	.097*	.158**		.156**
Productivity		.217**	272**	.157**
Organizing behavior	.150**			
Organizational politics	.092*			
Sensitivity	.149**	.100*	.136**	.085*
Decidability			.097*	
Flexibility	.191**	.169**	.143**	.128**
Team formation		.116**	202**	.116**
Communication	.151**	.180**		−.165**
Committed team leader		.111**		
N	266	266	266	266
R	.733	.747	.668	.748
R2	.538	.559	.447	.559
Adjusted R2	.516	.536	.423	.538
F	24.542	24.527	18.646	26.731
Probability>F	.000	.000	.000	.000

*$p<.05$, **$p<.01$

time, compatibility, an innovative and motivative approach, organizing behavior, organizational politics, sensitivity, flexibility, and communication. For the scope dimension, design and analysis of requirements, requirement selection, changeability, organizational policies, next-release development time, compatibility, innovative and motivative approach, productivity, sensitivity, flexibility, team formation, communication, and a committed team leader are significant among all the factors. The 11 success factors that are considered as significant for the time dimension are design and analysis of requirements, requirement selection, changeability, organizational policies, next-release development time, compatibility, productivity, sensitivity, decidability, flexibility, and team formation. Lastly, for the success of an NRP project in terms of cost, 12 factors are found to be significant: design and analysis of requirements, security of software, requirement selection, changeability, next-release development time, compatibility, an innovative and motivative approach, productivity, sensitivity, flexibility, team formation, and communication. As per the result, from the list of 68 hypotheses, only 48 were sustained, and the remaining 20 were excluded. Those hypotheses were excluded because of their high probability level and low coefficient values for their equivalent null hypotheses. Hence, it is shown that the existence of these factors did not make a significant contribution to the value of the success dimension. Table 9.5 encapsulates the results of hypotheses testing. The 48 hypotheses found to be supported are highlighted and labeled with a check mark (√), whereas the rest were the rejected hypotheses.

9.5.2 Discussion

The result obtained from the hypotheses testing and regression analysis supports us to answer the research question posed at the start of the research. We have calculated the success level of the combined four perceived success dimensions. Each success dimension carries its own perception of success, and the frequency of the supported hypotheses can be used to simplify the overall perception of success. The answers for all the research questions are discussed below:

Research Question 1: Are the identified factors actually the CSF of NRP-based software development?

From the findings, we can conclude that out of 17 factors, all have supported the hypotheses. The factors considered as the candidates of CSF are as follows:

Factor 1: Team formation (in terms of cost, scope, and time)
Factor 2: Communication (in terms of cost, quality, and scope)
Factor 3: Committed team leader (in terms of scope)

TABLE 9.5
Hypotheses Testing Results

Factors	*Quality*	*Scope*	*Time*	*Cost*
Team formation	H1a (√)	**H1b (√)**	**H1c (√)**	**H1d (√)**
Communication	H2a	**H2b (√)**	H2c	**H2d (√)**
Committed team leader	H3a (√)	**H3b (√)**	H3c	H3d
Organizing behavior	H4a (√)	H4b	H4c	H4d
Sensitivity	H5a	**H5b (√)**	**H5c (√)**	**H5d (√)**
Flexibility	H6a (√)	**H6b (√)**	**H6c (√)**	**H6d (√)**
Organizational policies	H7a (√)	**H7b (√)**	**H7c (√)**	H7d
Organizational politics	H8a (√)	H8b	H8c	H8d
Innovative and motivative approach	H9a (√)	**H9b (√)**	H9c	**H9d (√)**
Next release development time	H10a (√)	**H10b (√)**	**H10c (√)**	**H10d (√)**
Decidability	H11a	H11b	**H11c (√)**	H11d
Requirement selection	H12a (√)	**H12b (√)**	**H12c (√)**	**H12d (√)**
Design and analysis of requirements	H13a (√)	**H13b (√)**	**H13c (√)**	**H13d (√)**
Changeability	H14a (√)	**H14b (√)**	**H14c (√)**	**H14d (√)**
Productivity	H15a (√)	**H15b (√)**	**H15c (√)**	**H15d (√)**
Security of software	H16a	H16b	H16c	**H16d (√)**
Compatibility	H17a (√)	**H17b (√)**	**H17c (√)**	**H17d (√)**

Factor 4: Organizing behavior (in terms of quality)
Factor 5: Sensitivity (in terms of scope, time, quality, and cost)
Factor 6: Flexibility (in terms of scope, time, quality, and cost)
Factor 7: Organizational policies (in terms of time, quality, and scope)
Factor 8: Organizational politics (in terms of quality)
Factor 9: Innovative and motivative approach (in terms of cost, quality, and scope)
Factor 10: Next-release development time (in terms of scope, time, quality, and cost)
Factor 11: Decidability (in terms of time)
Factor 12: Requirement selection (in terms of scope, time, quality, and cost)
Factor 13: Design and analysis of requirements (in terms of scope, time, quality, and cost)
Factor 14: Changeability (in terms of scope, time, quality, and cost)
Factor 15: Productivity (in terms of cost, scope, and time)
Factor 16: Security of software (in terms of cost)
Factor 17: Compatibility (in terms of scope, time, quality, and cost)

Previous studies also supported the theory that factors like requirement selection, sensitivity, and next-release development time are the key factors for successful NRP-based implementation of software projects.

Research Question 2: How can we elaborate the relative significance of each factor while comparing them with the other factors?

On the basis of hypotheses testing, we can observe that design and analysis of requirements, compatibility, changeability, next-release development time, requirement selection, flexibility, and sensitivity had the most hypotheses supported (all four); this was followed by productivity, organizational policies, team formation, communication, innovative and motivative approach (three each); and finally, organizing behavior, security of software, decidability, organizational politics, and committed team leader (one). Ranking of each of the factors is given with respect to quality, scope, time, and cost in Table 9.6. For example, the impact of design and analysis of the requirement is ranked highest in terms of quality and cost, and hence is ranked one. Similarly, for scope and time, it is ranked as two and four, respectively.

Research Question 3: If significant difference is found amongst the absolute 17 CSF categorized on the basis of their effect on the success of NRP-based software project, can it be explained?

On the basis of regression analysis and hypotheses testing results, among the 17 categories, no significant difference was found. It is evident from the results that both tests have supported all 17 factors. Hence, all 17 factors have been accepted in both tests.

9.6 CONCLUSIONS

This chapter is a pioneering initiative in exploring and identifying CSF for the development of NRP-based software projects, which have meaningful implications for

TABLE 9.6
Ranking of CSF

Factor	Ranking Quality	Scope	Time	Cost	Frequency	Selection in the model
Design and analysis of requirements	1	2	4	1	4	.431, .308, .218, .416
Compatibility	3	4	3	3	4	.232, .248, .267, .219
Changeability	5	1	9	2	4	.314, .312, .137, .192
Next release development time	4	6	7	5	4	.179, .184, .159, .212
Requirement selection	2	10	6	6	4	.167, .141, .197, .354
Flexibility	10	8	8	4	4	.191, .169, .143, .128
Sensitivity	12	13	10	9	4	.149, .100, .136, .085
Productivity	7	5	2		3	.217, .272, .157
Organizational policies		3	1	10	3	.135, .288, .287
Team formation		11	5		3	.116, .202, .116
Communication	6	7		7	3	.151, .180, −.165
Innovative and motivative approach	8	9		11	3	.097, .158, .156
Organizing behavior			8		1	0.15
Security of software	9				1	0.133
Decidability			11		1	0.097
Organizational politics	12				1	0.092
Committed team leader		12			1	0.111

software professionals. Data was gathered from 266 software professionals based on judgmental sampling. Respondents were chosen from various organizations to gather the information for statistical analysis. Seventeen factors were identified through factor analysis and were organized into five different categories: people, organization, planning, implementation, and testing. A framework was proposed as an outcome of this analysis. A reliability test was then performed. The result of Cronbach's α test was 0.725, which is considered to be good. Since the CSF provides the direction for the success of the project, software professionals must therefore know what to focus on to improve performance and thus increase the bottom line while at the same time satisfying the clients for whom the software is being developed. It was found that out of 68 research hypotheses, only 48 were supported. The study proved that out of the 17 factors, seven are considered to be the most critical because of high frequency (i.e. factor 4). These factors are the design and analysis of requirements, compatibility, changeability, next-release development time, requirement selection, flexibility, and sensitivity. Design and analysis of requirement has the highest impact on quality and cost dimensions, as it is rank one in both dimensions, whereas the scope dimension is highly impacted by changeability. For the time dimension, the maximum impact is caused by organization policies. The key contribution of this research is to guide software professionals who are involved in NRP-based software projects. The study

TABLE 9.7
Result of Factor Analysis

Rotated Component Matrix

Questionnaire questions.	Component							
	Planning	Security	Requirement selection	Changeability	Organizational policies	Release time	Compatibility	Innovation and motivation
PEOPLE								
1								
2								
3								
4								
5								
6								
7								
8								
9								
10								
11								
12								
13								
Organization								
15					0.842			
16					0.651			
17					0.883			
18								
19								
20								0.673
21								0.563
22								0.727
Planning								
23								
24								
25								
26	0.854							
27	0.629							
28	0.619							
29	0.817							
30	0.816							
31						0.857		
32						0.844		
33								
34								
35								
36				0.875				
37				0.897				
38				0.533				
39				0.899				
40				0.512				
41								
42								
43								

Productivity	Organizing behavior	Organizational politics	Sensitivity	Decidability	Flexibility	Team Formation	Communication	Res-ponsibility	Com-munalities
									0.711
									0.787
	0.642								0.637
	0.717								0.757
							0.827		0.803
						0.676			0.761
						0.772			0.772
								0.779	0.753
			0.66						0.757
			0.599						0.724
					0.704				0.792
					0.636				0.687
									0.711
									0.869
									0.778
									0.863
									0.746
		0.74							0.692
									0.787
									0.757
									0.759
									0.661
									0.523
									0.665
									0.847
									0.645
									0.648
									0.787
									0.806
									0.893
									0.879
									0.656
				0.717					0.818
									0.655
									0.846
									0.879
									0.825
									0.886
									0.817
									0.759
									0.768
									0.782

(*continued*)

TABLE 9.7 (CONTINUED)
Result of Factor Analysis

Rotated Component Matrix

Questionnaire questions.	Planning	Security	Requirement selection	Changeability	Organizational policies	Release time	Compatibility	Innovation and motivation
44								
45								
Implementation								
46				0.843				
47				0.782				
48				0.842				
49				0.805				
50								
51								
52								
53								
54								
Testing								
55								
56							0.841	
57							0.64	
58								
59							0.656	
60								
61		0.918						
62		0.882						
63		0.906						
64		0.596						
65		0.893						
Eigen value	15.027	5.346	4.044	3.62	2.892	2.312	2.032	1.959
% Variance explained	9.434	6.982	6.759	5.697	5.342	4.851	4.432	4.312
Cumulative variance (%)	9.434	16.416	23.175	28.872	34.214	39.065	43.497	47.809

is also useful for researchers, as most of them are doing research in NRP evaluation and selection. This study would certainly help them to choose parameters for NRP evaluation and selection that are based on current industry practices.

9.7 LIMITATIONS AND FUTURE SCOPE

Though the present study has also added some new ideas to previous knowledge about the success of NRP-based software projects, it has several limitations as well. The first of these is the smaller sample size with respect to the population of software professionals involved in NRP-based software projects. For accurate and robust

Productivity	Organizing behavior	Organizational politics	Sensitivity	Decidability	Flexibility	Team Formation	Communication	Res-ponsibility	Com-munalities
									0.723
									0.571
									0.832
									0.764
									0.833
									0.772
									0.729
0.776									0.686
									0.738
									0.728
									0.733
									0.751
									0.823
									0.803
									0.673
									0.804
									0.713
									0.903
									0.878
									0.9
									0.69
									0.85
1.853	1.56	1.545	1.429	1.402	1.266	1.163	1.121	1.036	
4.057	3.658	3.347	3.069	3.032	2.996	2.917	2.872	2.56	
51.866	55.524	58.871	61.94	64.972	67.968	70.885	73.757	76.317	

statistical analyses, a large sample size is required. While many professionals were contacted during the e-mail campaign, the response rate of 12% was low. The second limitation is that not all dimensions of NRP software development were included in the study. Therefore, in future, more dimensions could be added to measure the success of software, for example, the stakeholders' perspective. The study would certainly present an encouraging contextual background to put forward a new conceptual framework for systematic analysis of an association between CSF and its related perspectives. Longitudinal study of this research can be carried out to identify which factors are critically significant in impacting the success of NRP-based software projects.

APPENDIX

The hypotheses formulated for the seven factors, each addressing four success dimensions, are presented as follows:

People
- H1: Team formulation is a CSF that enhances the success of NRP-based software projects in terms of (a) time, (b) cost, (c) quality, and (d) scope
- H2: Communication is a CSF that enhances the success of NRP-based software projects in terms of (a) time, (b) cost, (c) quality, and (d) scope
- H3: A committed team leader is a CSF that enhances the success of NRP-based software projects in terms of (a) time, (b) cost, (c) quality, and (d) scope
- H4: Organizing behavior is a CSF that enhances the success of NRP-based software projects in terms of (a) time, (b) cost, (c) quality, and (d) scope
- H5: Sensitivity is a CSF that enhances the success of NRP-based software projects in terms of (a) time, (b) cost, (c) quality, and (d) scope
- H6: Flexibility is a CSF that enhances the success of NRP-based software projects in terms of (a) time, (b) cost, (c) quality, and (d) scope

Organization Factor
- H7: Organizational policy is a CSF that enhances the success of NRP-based software projects in terms of (a) time, (b) cost, (c) quality, and (d) scope
- H8: Organizational politics is a CSF that enhances the success of NRP-based software projects in terms of (a) time, (b) cost, (c) quality, and (d) scope
- H9: An innovative and motivative approach is a CSF that enhances the success of NRP-based software projects in terms of (a) time, (b) cost, (c) quality, and (d) scope

Planning Factor
- H10: Next-release development time is a CSF that enhances the success of NRP-based software projects in terms of (a) time, (b) cost, (c) quality, and (d) scope
- H11: Decidability is a CSF that enhances the success of NRP-based software projects in terms of (a) time, (b) cost, (c) quality, and (d) scope
- H12: Requirement selection is a CSF that enhances the success of NRP-based software projects in terms of (a) time, (b) cost, (c) quality, and (d) scope
- H13: Design and analysis of requirements is a CSF that enhances the success of NRP-based software projects in terms of (a) time, (b) cost, (c) quality, and (d) scope

Implementation Factor
- H14: Corrective measures is a CSF that enhances the success of NRP-based software projects in terms of (a) time, (b) cost, (c) quality, and (d) scope
- H15: Productivity is a CSF that enhances the success of NRP-based software projects in terms of (a) time, (b) cost, (c) quality, and (d) scope

Testing Factor
H16: Security of software is a CSF that enhances the success of NRP-based software projects in terms of (a) time, (b) cost, (c) quality, and (d) scope

H17: Compatibility is a CSF that enhances the success of NRP-based software projects in terms of (a) time, (b) cost, (c) quality, and (d) scope

REFERENCES

1. Johnson, J. (2003), *CHAOS Chronicles v 3.0. Technical Report*. Standish Group International, Massachusetts.
2. Somerville, I. (2001), *Software Engineering* (6th ed.), Addison: Wesley Longman Publishing Co.
3. Bagnall, A., Rayward-Smith, V., and Whittley, I. (2001), "The next release problem", *Information and Software Technology, ELSEVIER*, 43(14), pp. 883–890.
4. Jones, C. (2006), "The economics of software maintenance in the twenty first century", *Ryan North and James Choi: Leveraging Software Performance Engineering to Enhance the Maintenance Process*, 68.
5. Sureka, A. (2014), "Requirements prioritization and next-release problem under non-additive value conditions", *Software Engineering Conference (ASWEC), 23rd Australian*, IEEE, Milsons Point, New South Wales, Australia, pp. 120–123.
6. Carlshamre, P., Sandahl, K., Lindvall, M., and Regnell, B. (2001), "An industrial survey of requirements interdependencies in software product release planning", *Requirements Engineering, 5th International Symposium*, IEEE, pp. 84–91.
7. Ruhe, G., and Moshood, O.S. (2005), "The art and science of software release planning", *Software, IEEE*, 22(6), pp. 47–53.
8. Zhang, Y., Harman, M., and Mansouri, S.A. (2007), "The multi-objective next release problem", *9th Annual Conference on Genetic and Evolutionary Computation*, ACM, pp. 1129–1137.
9. Roundtable, A. (2015), "The practice and future of release engineering", *Special Issue on Refactoring: Accelerating Software Change* 32(2), pp. 44–49.
10. Ebert, C. (2011), *Global Software and IT: A Guide to Distributed Development, Projects, and Outsourcing*. John Wiley and Sons. ISBN: 978-0-470-63619-0.
11. Ebert, C., Hoefner, G., and Mani, V.S. (2015), "What next? Advances in software-driven industries", *Software, IEEE*, 32(1), pp. 22–29.
12. Li, L., Harman, M., Letier, E., and Zhang, Y. (2014), "Robust next release problem: handling uncertainty during optimization", *Genetic and Evolutionary Computation, Conference Proceedings, ACM*, 7, pp. 1247–1254.
13. Larman, C., and Basili, V.R. (2003), "Iterative and incremental developments. a brief history", *Computer*, 36(6), pp.47–56.
14. Rosove, P.E. (1967), *Developing Computer-Based Information Systems*. John Wiley and Sons.
15. Aggarwal, K.K., and Singh, Y. (2009), *Software Engineering*. New Age Publications, New Delhi, India.
16. Leibowitz, J. (1999), "Information systems: Success or failure?" *Journal of Computer Information Systems* 40(1), pp. 17–26.
17. Reel, J.S. (1999), "Critical success factors in software projects", *IEEE Software*, 16(3), pp. 18–23.
18. Bytheway, A.J. (1999), "Successful software projects and how to achieve them", *IEEE Software*, 16(3), pp. 15–17.

19. Bosghossian, Z.J. (2002), "An investigation to the critical success factors of software development process, time and quality", Ph.D. thesis, Malibu, CA: Pepperdine University.
20. Chow, T., and Cao, D.B. (2008), "A survey of critical success factors in agile software projects", *The Journal of System and Software, Elsevier*, 81, pp. 961–971.
21. Nasir, M.H.N., and Sahibuddin, S. (2011), "Critical success factors for software projects: A comparative study", *Scientific Research and Essays*, 6(10), pp.2174–2186.
22. Zhou, X., Shi, Y., Deng, X., and Deng, Y. (2017), " D-DEMATEL: A new method to identify critical success factors in emergency management", *Safety Science*, 91, pp. 93–104.
23. Alrasheedi, M., and Capretz, L.F. (2015), "Determination of critical success factors affecting mobile learning: A meta-analysis approach", *TOJET: The Turkish Online Journal of Educational Technology*, 14(2), pp. 41–51.
24. Florén, H., Frishammar, J., Parida, V., and Wincent, J. (2018), "Critical success factors in early new product development: A review and a conceptual model", *International Entrepreneurship and Management Journal*, 14(2), pp. 411–427.
25. Durillo, J.J., Zhang, Y., Alba, E., Harman, M., and Nebro, A.J. (2011), "A study of the bi-objective next release problem", *Empirical Software Engineering, Springer*, 16(1), pp. 29–60.
26. Xuan, J., Jiang, H., Ren, Z., and Luo, Z. (2012), "Solving the large scale next release problem with a backbone-based multilevel algorithm", *Software Engineering, IEEE Transactions*, 38(5), pp. 1195–1212.
27. Del Sagrado, J., Del Águila, I.M., Orellana, F.J., and Túnez, S. (2010), "Requirements Selection: Knowledge based optimization techniques for solving the Next Release Problem", *6th Workshop on Knowledge Engineering and Software Engineering (KESE6)*, Karlsruhe, Germany, CEUR-WS.org, CEUR Workshop Proceedings, Vol. 636.
28. Colomo-Palacios, R., Tovar-Caro, E., García-Crespo, Á., and Gómez-Berbís, J.M. (2012), "Identifying technical competences of IT Professionals: The case of software engineers", *International Journal of Human Capital and Information Technology Professionals*, 1(1), pp. 31–43.
29. Veerapen, N., Ochoa, G., Harman, M., and Burke, E.K. (2015), "An integer linear programming approach to the single and bi-objective next release problem", *Information and Software Technology, ELSEVIER*, 65, pp. 1–13.
30. Van den Akker, M., Brinkkemper, S., Diepen, G., and Versendaal, J. (2005), "Determination of the next release of a software product: An approach using integer linear programming", *11th International Workshop on Requirements Engineering: Foundation for Software Quality (REFSQ)*, pp. 119–124.
31. Rubin, A., and Babbie, E. (1997), *Research Methods for Social Work*. 3rd edition. Pacific Grove, CA: Brooks/Cole Publishing Company.
32. Nunally, J. (1967), *Psychometric Theory*. New York: McGraw Hill.
33. Cronbach, L.J. (1951), "Coefficient alpha and the internal structure of tests", *Psychometrika*, 16(3), pp. 297–334.

10 Robotics and Machine Learning

Ghous Bakhsh Narejo

CONTENTS

10.1 Introduction: Robotics and Machine Learning Cooperation 184
10.2 Machine Learning Approaches in Robotics 184
    10.2.1 Computer Vision 184
    10.2.2 Imitation Learning 185
    10.2.3 Self-Supervised Learning 185
    10.2.4 Assistive and Medical Technologies 186
    10.2.5 Multi-Agent Learning 186
10.3 Research Trends in Robotics and Machine Learning 187
    10.3.1 Challenge 1: Complicated Multi-Dimensional Motion 187
    10.3.2 Challenge 2: Control Policies in Moving Landscape 187
    10.3.3 Challenge 3: Advanced Manipulation 188
    10.3.4 Challenge 4: Advanced Object Recognition 188
    10.3.5 Challenge 5: Assessment and Forecasting of Human Motion 188
    10.3.6 Challenge 6: Fusion of Sensors and Reduction in Dimensions 188
    10.3.7 Challenge 7: High-Rated Task Planning 188
10.4 Robotics, an Emerging Field 189
    10.4.1 Six Main Trends and Their Implications 189
        10.4.1.1 Commercial Investments 189
        10.4.1.2 Emergence of New International Players 189
        10.4.1.3 Reduction in Hardware Costs...................................... 189
        10.4.1.4 Popularity of Drones in the Civilian Sector 190
        10.4.1.5 Cloud Robotics 190
        10.4.1.6 Leveraging Social Media Data 190
10.5 Machine Learning Algorithms 190
    10.5.1 Types of Machine Learning Algorithms 190
        10.5.1.1 Supervised Learning Algorithms 190
        10.5.1.2 Unsupervised Learning Algorithms 191
        10.5.1.3 Reinforcement Learning 191
10.6 Data Science Helping Robotics 191
    10.6.1 Data Science and Robotics 192
10.7 Impact of Machine Learning on Robotic Technologies 193
    10.7.1 Machine Learning and Its Applications in Robotics 193
        10.7.1.1 Computer Vision 193
        10.7.1.2 Imitation Learning 193

10.7.1.3 Assistive and Medical Technologies 194
10.8 Learning Involving the Multi-Agent 194
10.8.1 Challenges 195
10.9 Conclusion 197
Bibliography 197

10.1 INTRODUCTION: ROBOTICS AND MACHINE LEARNING COOPERATION

Robotics can be described as an interdisciplinary branch of engineering that may include mechanical, electronics, information, computer, and other engineering faculties. The main purpose of robotics is to deal with the design, construction, operation, and use of robots as well as computer devices for their control, feedback, and the processing of data.

These innovations are used to create machines that can substitute for people and reproduce human activities. Robots can be utilized in numerous circumstances and for many purposes, but nowadays many operate in perilous conditions, in manufacturing processes, or where human beings cannot survive.

Nowadays, robotics is a highly progressive field, as the technology is advancing; the design of robots serves many different purposes, whether locally, commercially, or militarily. Numerous robots are built to do jobs that are dangerous for human beings, such as defusing bombs, finding survivors in unstable ruins, and investigating mines and wrecks. Robotics is additionally utilized in STEM (science, technology, engineering, and mathematics) as an instructing aid. The development of nano robots—microscopic robots that can be infused into the human body—can revolutionize medicine and human health.

Machine learning (ML) is a study based on the use of algorithms and models involving statistical analysis that computer systems utilize to perform a particular task without an open set of instructions but rather a specific set of patterns and inference. ML is also considered a sub-branch of artificial intelligence. The algorithms in ML build a model which is dependent upon a data set called the *training set*, and the predictions are built upon the same. These algorithms are employed in a range of applications like filtering emails and computer vision, especially in environments where conventional algorithms are not easy to operate.

ML, being dependent on statistics, is computed through the machines which is basically used to forecast trends based upon the data set utilized. Data mining, on the other hand, relies on unsupervised learning. In its application to business problems, ML is also referred to as *predictive analytics*. ML finds applications both now and in the future in multiple areas to do with robotics technologies.

10.2 MACHINE LEARNING APPROACHES IN ROBOTICS

10.2.1 COMPUTER VISION

Machine vision or robot vision, also known as *computer vision*, is based upon computer algorithms, as robotics experts use a camera for robotic vision to capture

Robotics and Machine Learning 185

physical data. The robot vision is coupled with machine vision, assisting the former to boost the robot-assisted human guidance and machine prediction. The small differences between the two terms may be ignored as robotics involves the adjustment of the frame of reference and the capacity of the robot to physically alter its surroundings.

There is a trend toward big data based on visual data powered by the internet, which consists of photos as well as videos which are marked or labelled. This trend has advanced computer vision based upon ML algorithms which predict the new techniques at the world class universities such as Carnegie Mellon and in the rest of the world, influencing robotics in applications like sorting and locating objects. There is one branch which focuses on detecting abnormalities in silicon wafers using unsupervised learning. These techniques use the convolution neural networks developed at the Biomimetic Robotics and Machine Learning lab in Munich. The robotics technologies consist in applications in such areas as radar, lidar, and ultrasound, which are supported by Nvidia, focused on vision-based systems for the control of autonomous vehicles and drones.

10.2.2 IMITATION LEARNING

Imitation learning is akin to the learning based on observation which is common in infants and toddlers. This is also an umbrella term that covers the scope of reinforcement learning. The latter is defined as an act by an agent in the world who works in order to maximize rewards as a challenge of attainment. It is an old research question related to research in the area of human-like robots, which is the closest that machine learning can get to the way humans learn. In reinforcement learning, computers and robots are taught to perform particular tasks in relation to their environment to create results that get either punishments or rewards. Hence, each time robots can learn from their mistakes in order to know what course of activity is required to achieve a reward.

Imitation learning has become an essential element of learning in the field of robotics, especially in areas whose working environments have an outdoor physical setting, such as construction, agriculture, search and rescue, and the military. It is a big challenge for humans to program these robotic solutions. Typical examples include the inverse optimal control methods, or "programming by demonstration," which are applied in robotic applications.

Another technique that has been popular in the field is the Bayesian belief network, which has been tested on forward learning models, making it easy for a robot to have a priori knowledge of its motor behavior under the influence of the external environment. The motor babbling (motor babbling is a process of repeatedly performing a random motor command for a short duration) has been tested on a robot at the University of Illinois at Urbana-Champaign (UIUC).

10.2.3 SELF-SUPERVISED LEARNING

Self-supervised learning assists robots in creating situations to improve performance. It consists of prior or advance knowledge. The data is scrutinized closely to detect large datasets. It is utilized in robots to detect and block dust and snow and to identify vegetation and obstacles in uneven terrain using the vision-based analysis used in the modeling of transport vehicles.

Robots designed by researchers employ sensors to track humans as per patterns learned through probability. A laser pointer is used to aim at the object, such as spilled milk. It is successful in 60 percent of cases as the researchers increased these trials. The method has also been applied to a traffic flow and road detection algorithm using a front-view monocular camera at MIT applied to mobile on-road robots. Autonomous learning finds applications in robot and other control tasks. Researchers have devised a technique to enhance learning using the model of uncertainty, using it in a long-term planning and controller-based learning, which has improved in terms of reducing the effects of model errors while facilitating the robots to learn a new skill.

10.2.4 Assistive and Medical Technologies

The assistive robot is a device that processes sensory information and performs actions which benefit humans suffering from disabilities and old age, assisted by smart assistive technologies which also exist for the betterment of the general population—for example, driver assistance tools. Movement therapy robots can also have a diagnostic or therapeutic effect that is of benefit to the general public. All these are technologies are largely limited in their scope as they are mostly restricted to the lab, being costly to be afforded by most hospitals in the United States and the rest of the world.

A desktop vocational assistant robot (DeVAR) was designed by researchers at Stanford in the 1990s along with Palo Alto Veterans Affairs Rehabilitation Research and Development. Machine learning–based robotic assistive technologies have also been developed recently which use a combination of assistive machines exhibiting autonomous behavior. The MICO robotic arm, developed at Northwestern University, senses the environment using a Kinect Sensor. The implications of these new innovations are quite complex, and while these smarter assistive robots easily adapt to the needs of the users, they are still partially autonomous.

Progress in the machine-learning techniques concerned with the medical field with applications in robotics has been made rapidly, notwithstanding the slow progress in the implementations in the field. One example is the research cooperation between the Centre for Automation and Learning for Medical Robotics, various universities, and a cluster of physicians towards the formation of the 'Smart Tissue Autonomous Robot (STAR)' at Children's National Health System, DC. The progress in medical robotics, especially in 3D sensing and the STAR project, has resulted in the stitching of pig intestines showing greater precision and reliability as compared with surgery by humans. Experts caution, however, that STAR will not replace human surgeons, who will be controlling the surgery while robots will assist them. This, however, will help the surgeons in performing delicate surgical procedures.

10.2.5 Multi-Agent Learning

In multi-agent learning, a coordinated and cooperative environment based upon machine learning–based robots or agents is also used extensively in games, where the agents are capable of adaptation to the ever-changing environment, resulting in strategies which are called *equilibrium* strategies.

A better and strong case involves an algorithm focusing on robots located in a cooperative environment devised for information and decision systems. In this design, the robots collaborated with each other in order to build a comprehensive learning model better than single robot, using a smaller set of data processed together; the concept is based upon exploring a building and its room layouts and autonomously constructing a knowledge base.

In this experiment, a robot constructed its own catalog, and worked towards a collective effort with other data sets on robots. The resulting distributed algorithm also excelled the reference algorithm using this knowledge base. The system is relatively perfect, and in the design involving machine learning, robots must refer to the data sets in order to strengthen their mutual observations. This results in mistakes being corrected, which will definitely play a futuristic role in a diverse set of robotic applications.

10.3 RESEARCH TRENDS IN ROBOTICS AND MACHINE LEARNING

Deep learning is a science which involves the training of large artificial neural networks. These are also called deep neural networks (DNNs), consisting of so many parameters which assist to model complex functions like nonlinear dynamics. Researchers in the field of robotics have narrowed down various goals for robotics over the next two decades. These are walking and running imitating a human, teaching through practical examples, mobile navigation alongside roads, collaboration-assisted automation, self-picking of objects, recovery without manual intervention, automatic aircraft inspection and maintenance, and reduction and healing of robotic damage. There are, however, some serious challenges for robotics that DNN technology can help to address.

10.3.1 CHALLENGE 1: COMPLICATED MULTI-DIMENSIONAL MOTION

The complex dynamics involving analytical offshoots need human expertise; this process is very detailed and takes a lot of time, and it also means a compromise between state dimensionality and tractability. However, these workable models which are robust and without uncertainty are difficult to realize and their complete data set is lacking. These systems must quickly and autonomously adjust to unknown and new dynamics which are needed to solve problems such as grasping new objects, traveling over surfaces with unknown or uncertain properties, managing interactions between a new tool and the environment, or adapting to degradation and/or failure of robot subsystems. Also required are the most desired methods for completing tasks with multiple DOF with high uncertainty and less information.

10.3.2 CHALLENGE 2: CONTROL POLICIES IN MOVING LANDSCAPE

In cases involving moving systems which can deal with large degrees of freedom in application sets like multiple limbs and moving manipulators, anthropomorphic limbs, and swarm robotics are a must. Robots will be employed to work as required in situations which contain uncertainty along with limited information on the state space.

10.3.3 Challenge 3: Advanced Manipulation

Despite the fact that advances have been made over last 30 years due to intensive research, reliable solutions in the case of jobs such as holding a variable shape as well as complicated structures, making use of tools, and the actuation systems involved in a context like turning a valve, opening a door, and so on remain a quite unclear set of tasks, especially when taking into account novel situations. This challenge includes the kinematic, kinetic, and grasp planning inherent in tasks such as these.

10.3.4 Challenge 4: Advanced Object Recognition

DNNs are set to remain an expert at recognition and classification of objects. Examples involving advanced applications are resetting the detection and estimation of varying shapes of objects, estimating the environment, and assessing the path, as well as recognizing acts of motion in a particular way—for example, any small tasks of motion like moving around the table, opening or closing a car door or trunk—and detecting rough or uneven surfaces which all are not human friendly.

10.3.5 Challenge 5: Assessment and Forecasting of Human Motion

This is an important challenge in robots which are required to work in the midst of people in areas of collaborative robotic environments used in applications such as manufacturing, care of the elderly, designing autonomous vehicles that work at crowded road crossings, and roadside walking tracks for individuals. The processes encourage learning through practical examples, and this will help those who are not experts in these fields and lack knowledge of robotics and programming. The same challenge is extendible to human needs when human perception needs to be detected and the need for manual intervention is required during the process.

10.3.6 Challenge 6: Fusion of Sensors and Reduction in Dimensions

The development of economical sensing-based technologies is quite popular in robotics, creating a variety of possibly rich, high-dimensional, and multimodal data sets. The challenging methods focus on developing fruitful representations of state from such data sets.

10.3.7 Challenge 7: High-Rated Task Planning

The reliable execution of high-level tasks can be added to the previous challenges required to be attained by robots to achieve a higher level of utilization for cases involving general public benefits. General commands involve lower-levels tasks like getting milk from the refrigerator: closing or opening the door, identifying the right

Robotics and Machine Learning

container, and grasping the container in a secure way. These tasks are based upon challenges 1 and 2, and the objectives of teaching through practical examples also need modernization, as discussed above in challenges 4–6.

10.4 ROBOTICS, AN EMERGING FIELD

There are signs all around us showing that the field of robotics is going through a major change. Robots are getting noteworthy coverage within the media. A number of huge companies that had little to do with robotics are all of a sudden on a buying spree to procure robot companies. Nations that were not on anyone's radar screen just a few years ago are presently emerging as major players within the robotics field. Costs relating to new applications are falling quickly. Even the idea of what was considered a robot is rapidly changing. All these signs appear to indicate that robotics is on the threshold of something huge that could ideally affect our lives in a positive way.

10.4.1 Six Main Trends and Their Implications

10.4.1.1 Commercial Investments

As of late, the commercial sector has made critical investments in the field of robotics. A few robotics companies have been bought by Google, Amazon has purchased Kiva Systems and morphed it into Amazon Robotics, and Qualcomm has also made investment in the field of robotics. The Toyota company is the world's largest car manufacturer, and it is reported that it is also entering into the field of artificial intelligence and robotics. Even venture capitalists are now taking interest in funding robotics businesses. Optimistically, this will lead to the implementation of robotics in new applications and will accelerate innovation in the technological field.

10.4.1.2 Emergence of New International Players

Traditionally, advances in robotics came for the most part from America, Japan, China, and European countries. Nowadays, the robotics field is extending and new global players are emerging. China is playing a vital role in investments in the robotics field. Today, Chinese producers are leading the world with new industrial robots by producing low-cost robots. Chinese firm DJI is the world's largest manufacturer of commercial drones. Recently, South Korea has also been emerging in the field of robotics. A group from South Korea built a robot that won the competition in the DARPA robotics challenge, competing against teams from America, Japan, and Europe. Countries like the Netherlands, Switzerland, the United Arab Emirates, and many others are contributing intensely to the fields of robotics, artificial intelligence, and drone technology. New opportunities are also anticipated due to the globalization of the robotics field.

10.4.1.3 Reduction in Hardware Costs

The cost of robots like mobile robotic platforms, articulated manipulators, and drones have been declining in the commercial sector. This is predicted to enable the use of robots in new applications. The agricultural sector could also be a major new market for robots and unmanned aerial vehicles (UAVs).

10.4.1.4 Popularity of Drones in the Civilian Sector

Drones play a vital role in the civilian sector both locally and globally, one that is anticipated to grow at a rapid rate. Unfortunately, these robots are vulnerable from a cyber-security point of view. For example, hacking of cars shows the defenselessness of these vehicles to cyber-attacks. Nowadays, modern cyber-security innovations are required to deal with cyber-attacks which can commandeer vehicles and cause physical harm. Public opinion could be impacted by a serious incident in this region, which could cause a major difficulty for this emerging field.

10.4.1.5 Cloud Robotics

Robots can use clouds to do substantial data processing and interchange data with other robots in real time. With the help of cloud robotics, robots can be freed from computing limitations and be given "big enough brains" to deal with challenging circumstances which were not possible for them to deal with earlier. The robotics community is embracing the advances in "big data" to deal with the huge amounts of data produced by sensor-rich robots.

10.4.1.6 Leveraging Social Media Data

Today data (like pictures, maps, and videos) can be accessed by robots on social media. Deep learning algorithms are artificial intelligence (AI) techniques, which are used to create new perception capabilities that can be picked up in order to extend AI's capacity to understand the environment. Robots are securing new skills by using social media.

10.5 MACHINE LEARNING ALGORITHMS

Machine learning algorithms can learn from data and improve from involvement without any human intervention. There are several learning tasks which may include learning the function that maps the input to the output; learning the covered up structure in unlabeled data; or "instance-based learning", where a class label is created for a modern occurrence by comparing the unused instances (row) to instances from the training data, which were stored in memory. "Instance-based learning" does not make a abstraction from particular instances.

10.5.1 Types of Machine Learning Algorithms

There are three types of machine learning (ML) algorithms as follows.

10.5.1.1 Supervised Learning Algorithms

Supervised learning uses labeled training data to memorize and learn the mapping function that turns input (X) into output (Y). In other words, it solves for f within the equation:

$$Y = f(X)$$

Robotics and Machine Learning 191

This permits us to precisely and accurately generate output when given new inputs. There are two types of supervised learning: *classification* and *regression*.

- *Classification* is a type of supervised learning which predicts the results of given input sample when the output is in the shape of classification. A classification model might see the input data and attempt to anticipate names like "sick" or "healthy."
- *Regression* is another type of supervised learning which predicts the results of given input sample when the output is in the shape of real values. For illustration, a relapse exhibit might handle input data to anticipate the sum of precipitation like rainfall or a person's height and so on.

10.5.1.2 Unsupervised Learning Algorithms

Unsupervised learning algorithms models are used when we have the input (X) and no resultant output (Y). In these models, unlabeled training data is used to model the basic structure of the data.

There are three types of unsupervised learning:

- *Association* is a type of unsupervised learning used to find the possibility of co-occurrence of items in a collection. It is broadly used for analysis of market-basket data. For illustration, an association model could be utilized to find that on the off chance that a client buys bread, s/he is 80% likely to buy eggs as well.
- *Clustering* is used to gather samples such that items inside the same cluster are more analogous to each other than to items from another cluster.
- *Dimensionality reduction* is used to decrease the variables of a data set while ensuring that critical data is sent. Feature extraction methods and feature selection methods are used in dimensionality reduction. Feature selection is used to select a subset of the initial variables. Meanwhile Feature extraction moves the data from a high dimensional space to a low dimensional space.

10.5.1.3 Reinforcement Learning

Another type of machine learning algorithm is known as reinforcement learning; this permits an agent to choose the next activity based on its current condition by learning activities that will exploit a reward. Reinforcement algorithms are used to learn ideal activities via a trial and error method. For example, a player wants to move to certain places at certain time in a video game to win. The game will begin by moving randomly over time through trial and error in order to learn.

10.6 DATA SCIENCE HELPING ROBOTICS

In the digital revolution, big data and data science play a vital role in technologies like machine learning, artificial intelligence, and deep learning. The pith of data science is to jump into substantial datasets to extricate important data from them.

Various institutions and organizations over different divisions of the industry are presently leveraging data science advances to control development and technology-driven change.

As increasing numbers of companies are moving towards data science to improve their organizational foundations, this is giving rise to exciting career openings such as data architects, data researchers, data analysts, machine learning engineers, and so on. Hence, if you want to begin a career in data science, the most fruitful time is now. Today, there are a lot of resources accessible to assist you to get started in the field of data science.

10.6.1 Data Science and Robotics

The field of robotics has certainly made great strides with progress in data science. During the early days of development, researchers and scientists were confronted with two major challenges, one being to predict the action of a robot and the second to reduce computational complexity in real time. However, robots may perform particular functions in which it is impossible for researchers to foresee their next move. A robot would have to be reprogrammed for every new functionality at every time, which would be a tedious task. Another complication with robots is that unlike people, who use their matchless sense of vision to form a sense of the world around them, robots can only imagine in arrangements of zeros and ones. In this way, robots would have to generate a set of zeros and ones for fulfilling real-time vision tasks every time an unanticipated change develops, thereby expanding the computational complexity.

These issues in robotics can be solved by machine learning. Robots can establish new behavior patterns via labeled data with machine learning. For example, in handwriting recognition, both positive and negative are labeled data that are fed into computers. Once the computer has effectively learned, it can easily distinguish between positive and negative and display this ability with new data. During the training stage, the computer can foresee the qualified classifiers for recognizing the handwriting. With the progress of machine learning algorithms, computers are now able to perform handwriting recognition much more precisely than they were ten years ago.

Data researchers and data analysts continue to leverage machine learning and artificial intelligence to create smart machines. During the process, they pick up a deeper insight into the world of data science. Data researchers, scientists, and analysts can handle, analyze, and understand vast data sets much more quickly than ever. For instance, the MIT data science machine can handle large amounts of data, whereas the same data would take months to process if it were to be done physically by data scientist and analysts.

In this way, robotics, artificial intelligence, and data science have a beneficial mutual relationship. Each boosts the other to control innovative machines that are making our lives easier than ever. The collaboration of data science, machine learning, and artificial intelligence has given us helpful things like smart assistants, robo-surgeons and nurses, and self-driving cars.

10.7 IMPACT OF MACHINE LEARNING ON ROBOTIC TECHNOLOGIES

Nowadays, machine learning technologies are influencing the use of robotics in many innovative technological fields. Recently, Evans Data Corporation Global Development published the results of a survey showing that robotics and machine learning is at the top of the priorities of 56.4% of developers, and about 24.7% of all developers are engaging in building robotics apps and using machine learning in their projects.

There are five key areas where machine learning has played a vital role in the field of robotics technology. Currently machine learning and robotics both have important applications in future development stages.

10.7.1 Machine Learning and Its Applications in Robotics

10.7.1.1 Computer Vision

Rather than computer vision, this may be called *machine vision* or *robot vision*, which involves a camera and other hardware for which robots are required to automate the data processing. The credit for the emergence of robot guidance and automatic inspection systems is due to the close link between robot vision and machine vision. The difference between them may be due to motion as utilized in visual processing by robots, with scope in optimizing the capability of machines to alter the environment physically.

An entry of large data, that is, visible data assisted by the internet, has boosted advances in computer vision, further assisting the development of machine learning based on prediction of learning techniques at universities and other institutions, foremost in areas such as robot vision, which is used for the identification and sorting of objects of interest. For example, unsupervised learning exhibited anomaly detection, such as when a building system has the ability to find and assess faults in silicon wafers using a convolution neural network at the research labs concerned.

10.7.1.2 Imitation Learning

Imitation learning is a method of learning through observation, mimicking the behavior of newborns and children. Imitation is a category of reinforcement learning. Models like Bayesian or probabilistic techniques are common functions of this machine learning approach.

Imitation learning has become an essential part of the robotics field, to perform a task from demonstrations by learning a mapping between observations and actions. For example, CMU and other organizations have applied the techniques in inverse optimal control methods using programming tools. Bayesian theory has been applied to learning models, in which a robot learns without a priori knowledge of its motor system or external environment. Robots are capable of generating their own training with the approach of self-supervised learning, for example to improve their performance; beforehand, field-based workshops imparted learning by analyzing the data, which needed to be captured in order to interpret the situations shown in the long-rang sensor data. These conditions needed the data to be embedded into robots and other sensors that are used to block objects and obstacles in rough terrain environments using the scene analysis and modeling of moving vehicles.

A real robot machine has been created by researchers using a Kinect sensor, a camera, a computer, and a laser pointer for observation of human movement. These are patterns that it learns on the basis of probability. A laser pointer is used by Watch-Bot to identify and remember the object. In preliminary tests, the trials were extended by researchers in order to allow the robot to learn from online videos, and bot was capable of recognizing humans the majority of the time.

In robotics, other examples of self-supervised learning methods consist of an algorithm using road data for detecting using a front-view camera with a road probabilistic distribution model and fuzzy support vector machine.

Independent learning is different from self-supervised learning, in which deep learning and unsupervised methods are included in order to utilize robots and control tasks. A team of researchers from Washington and Cambridge universities at Imperial College in London has created a new method for rapid learning that integrates a probabilistic model into long-term control learning and planning and decreases the effect of model error when learning a new skill.

10.7.1.3 Assistive and Medical Technologies

An assistive robot is type of device that can not only sense but also process data from a sensor and carry out actions that benefit people with disabilities. Movement therapy robots are beneficial for both diagnostic and therapeutic purposes. Unfortunately, these both technologies are still confined to the lab, as they are still cost-prohibitive for most hospitals.

In the 1990s, assistive technologies were exemplified by the robot developed by Stanford and the organization dealing with veterans at Palo Alto. The other cases where deep learning is being used in robotics are combined assistive machines with autonomy, like the MICO robotic arm that observes the world through a 3D sensor (Kinect).

Progress involving the methods in the machine learning used in robotics is fast enough and is rapidly evolving.

10.8 LEARNING INVOLVING THE MULTI-AGENT

Multi-agent learning has important components like coordination and negotiation, which are involved in machine learning–based robots.

In late 2014, researchers from MIT have created an algorithm for distributed agents or robots for information and decision systems. The collaboration of robots built a much more complete model of learning than could be done with one robot, based on knowledge of autonomy and room layout using the concept of exploring a building.

Robots individually built their own sequences involving other robots through the data used in the distributed algorithm, and outclassed the reference algorithm in developing these sequences on the basis of information. The machine learning methodology being discussed here permits robots to distinguish catalogs and data sets,

Robotics and Machine Learning

emphasize mutually sensed data, and edit the mistakes in the data, thereby increasing their potential role in a diverse set of areas in the future, including airborne vehicles.

10.8.1 Challenges

Robotic systems consist of software and hardware components whose characteristics change as time passes. All the design parameters assumed so far do not hold true at implementation stage. There is a technique called self-configurability which enables robot systems to adapt to such inconsistencies which arise at run-time, although this does not hold true for systems where these techniques can't be planned due to the large scope of the task. However, these challenges may be overcome by using machine learning to find Pareto-optimal configurations by reducing the search space and making the planning task handy and doable.

The parameters involved in this technique are task timelines and energy consumption. The technique provides a plan that ensures a high-quality adaptation to unknown, probable, and harsh surroundings.

In modern system design, design components have the ability to modify their behavior as time passes. The quality of internet services degrades as it is affected by implementation flaws, and this may also be true for other cyber-physical systems where stage-wise progress in reliable operation and decrease in physical damage is desirable.

Software fails to continue to work in all such situations, negating the initial assumptions of the design constituents used in a system. Self-adaptive systems are no exception as progressive degradation also mars their expected performance.

Software is the right platform for studying the failure of self-adaptive mechanisms, as assumptions about their system constituents break due to their weak nature. Sensors degrade over time, becoming inaccurate and suffering from unexpectedly consuming more energy. Self-adaptive system models are designed manually by domain experts while designs suffer from flaws such as being expensive and potentially unreliable or too small to be adjusted as the system evolves.

Independent of any approach adopted, the large scope of configurations and surrounding conditions almost invariably result in simplistic assumptions about the working models which fail to reflect accurately the critical interdependency between the constituents, the options on configuration, and the external variables.

Let us take the example of robot models which provide information about power consumption. The models are built in order to assist the robot in adapting to new environments in order to save power. These models may also help the robot in determining environmental uncertainties and other limitations. The robot may adapt to a new environment by using a less accurate but more efficient algorithm which helps in increasing the battery life for the rest of the mission time. This algorithm will be very productive if the underlying assumption on which the decision making is based is reliable. Therefore, the estimation of energy must be accurate in order to expect accurate assessment and saving of the battery life.

In cases where the energy demand goes higher than the design expectation, the robot mission will fail to complete. In that case, a software based configuration helps the robot to choose to optimize the usage of sensors when the need arises.

An appropriate configuration for a self-adaptive system needs a viable plan designed to function optimally if adapted to the configuration at run time, a key challenge as the configuration space is exponentially large, consisting of millions or even billions of possible configurations. The planning is challenging as the large size of the planning space means that the system to be adopted needs to be highly exhaustive.

The optimal performance of the planning problem may be dealt with in a better way through its run-time adaptation, using machine learning to help in finding the Pareto-optimal configurations that can achieve the lowest energy consumption as well as better timeliness.

The elements of the learning and quantitative planning must go hand in hand to make the run-time self-adaptation possible, and the data obtained from many heterogeneous models must be made available in quantitative planning in order to attain the best energy consumption and timeliness and their interdependencies.

Mobile robotic systems are complex combinations of elements such as physical hardware and robotic control software which helps them to sense and move through an external landscape, and they are increasingly relied on in modern society, giving the assistance in areas such as healthcare, the transportation of objects and logistics, environmental protection from hazards, and assessment and maintenance of infrastructure suffering from external dangers.

In all such areas, these machines perform a list of tasks based upon parameters such as long-range, long-term, and complex goals.

Energy management of mobile robots needs to be accurate in order to avoid mishaps and hazards like the possibility of running out of power in the middle of an operation. This is not true in the case of industrial robots, however, as energy is assumed to be available throughout the operation time. Data regarding the characteristics of the mission, the operating environment conditions, and other diverse robot configurations covering sensors, actuators, computation intensive control algorithms, and more show high variation, making it very difficult to predict energy consumption.

The goal of the robot in such situations is to cover the maximum area with the minimum energy consumption and to attain the goal within minimum amount of time within battery limitations. During this process, the robot must face an unpredictable environment and events based upon the obstacles blocking its path, and abrupt variations in battery energy level due to the sensor, malfunctions, or rough terrain making it hard to navigate. The robot can adapt to such situations in order to avoid the occurrence of the failures in the mission by reconfiguring to a reduced energy consumption mode and changing the direction and the path to attainment.

The objective is to attain self-adaptation for systems that can be configurable as these systems work under constant uncertainty. The configuration helps the systems to adapt to uncertain environments. The challenge of large space is overcome by finding the most critical configurations. These configurations are the most important, as they limit the explorations. The most optimized configurations should be sought in order to obtain optimum results.

The key insight of our approach is that we reduce the adaptation search space, in the offline phase, by incorporating information about the optimal configurations into the models that the adaptation layer employs for planning in the online phase. The reduction in the adaptation workspace achieved through the optimized configuration holds the answer to the challenge. The system output parameters are measured in terms of more than one aim, which are the localization errors and the central processing unit (CPU) utilization of power or energy.

If system configurations are binary, then if the option includes the performance model, then it is influential. All other options in numeric and categorical terms may also be transformed into binary by choosing these to be zero or one, while compromising the accuracy at the same time. Due to the existence of the term in the statistical analysis, the results on this model provide the deciding options and interactions. The learning consists of several steps: initialization, forward selection, backward elimination, and termination. The final result consists of parameters that are quite important for the study.

10.9 CONCLUSION

Robotics and machine learning have traditionally and historically remained separate until recently, when both fields have developed inroads into each other. Advancements in machine learning have resulted in deep learning, and this has benefitted robotics. Robotics is an emerging field encompassing multiple trends and their impacts: commercial investments, introduction of global competitors, reduction in hardware costs, popularity of drones in the civilian sector, cloud robotics, and leveraging social media data. Machine learning algorithms have a huge impact as these algorithms have been applied in robotics for supervised as well as unsupervised learning. This, aided by the data science, has helped robotics in a huge way.

BIBLIOGRAPHY

Daniel Faggella. *Machine Learning in Robotics: 5 Modern Applications.*
Harry A. Pierson, Michael S. Gashler. Deep Learning in Robotics: A Review of Recent Research. *Advanced Robotics*, 2017. *31*(16), 821–835.
Bradley Schmerl, Christian Kastner, David Garlan. (2019). Machine learning meets quantitative planning: Enabling self-adaptation in autonomous robots. In *2019 IEEE/ACM 14th International Symposium on Software Engineering for Adaptive and Self-Managing Systems (SEAMS)* (pp. 39–50). IEEE.

11 Detecting Medical Reviews Using Sentiment Analysis

*Sandhya Makkar, Mayank Singhal,
Nimish Gulati, and Shivani Agarwal*

CONTENTS

11.1	Introduction	199
11.2	Background	200
11.3	Data Transformation	201
11.4	Filtering	201
11.5	Lexicon-Based Method of Sentiment Analysis: Lexicon Construction	201
11.6	Classification	203
11.7	Evaluation	204
11.8	A Review of Sentiment Analysis	206
11.9	Performing Sentiment Analysis on Medical Reviews	209
11.10	Sentiment Analysis Working Program	212
11.11	Conclusion	214
Bibliography		216

11.1 INTRODUCTION

Sentiment analysis of reviews is the process of analyzing product reviews available online to determine the overall opinion or feeling about a product or service. Reviews represent people's opinions and express how they feel regarding the product or service. These reviews are receiving increasing attention and are a very useful resource for marketing teams and other statistical teams who want to dig deep into public opinions about products and services. Sentiment analysis provides deep insight into these opinions, which are of great importance for industry analysts.

The huge amount of reviews present on the web represent the opinions of users. It is hard for humans to summarize and generalize this large amount of data due to its diversity and size, and this has led to the need for automated and real-time opinion extraction and mining. Deciding about the sentiment of opinion is a difficult task as it largely depends on the subjectivity factor of the content, which is largely what people think.

Sentiment analysis is a method of classification as it classifies the polarity of the text as either positive, negative, or neutral. Machine learning is used as the tool in sentiment analysis for classification in addition to lexicon based approaches. It has been suggested by researchers that machine learning classifiers such as naive Bayes, maximum entropy, and support vector machine (SVM) provide 90% accuracy in classification in case of topic categorization, but in the case of sentiment analysis, this accuracy falls to 78%–85% due to the nature of the text that requires better understanding.

11.2 BACKGROUND

There have been many studies carried out which deal with different levels of analyzed text: word or phrase level, in which the orientation of the words or phrase effect the overall orientation of the text; sentence level, in which the whole sentence is considered to express the overall orientation of the opinion; and document level, in which the whole document is used to defining the orientation of the document. There are three approaches in sentiment analysis, which can be classified as the machine learning based approach, the linguistic analysis approach, and the lexicon-based approach. Machine learning methods are based on training the algorithm to carry out classification based on selected features for a specific task, testing this algorithm on different sets of texts, and checking whether is comes up with the correct features and classification. In lexicon-based methods, a predefined collection of words with their polarities are used, and then an algorithm is used to find their occurrences and assign weights and provide the overall polarity to the text. In the linguistic approach, the syntactic characters of the words or phrases and the negation and structure of the text are used to determine the orientation of the text. This is usually combined with the lexicon-based approach.

According to Haddi *et al.* (2013), pre-processing is the process in which the text is cleaned and prepared for classification. Online text usually contains much unwanted noise like HTML tags, scripts, and ads, and it needs to be cleaned of these before it can be used. In addition, many words have no impact on the orientation of the text. These words need to be removed as these words increase the dimensionality, because each word adds a dimension. Reducing noise from the text helps improve the performance of the classification process, which helps in speeding up sentiment analysis.

The process of sentiment analysis involves the following steps: text cleaning, extra white space removal, expanding abbreviations, stemming, removal of stop words, handling of negation, and selection of features. All the steps except the last one are called *transformations*, while the last step, in which some functions are applied to select required pattern, is called *filtering*.

Features are words and phrases in the context of sentiment analysis which strongly express the opinions as positive, negative, or neutral. They have higher impact on the sentiment of the text as compared to other words in the same text. There are several methods for feature selection; for example, some are based on the syntactic position of the word such as adjectives; some are univariate, based on the feature's relation to a specific category, such as chi squared and information gain; and some are multivariate, using genetic algorithms and decision trees. The method to assess the importance of each feature is by assigning weights to each feature. The most popular methods are feature frequency (FF), term frequency (TF-IDF), and feature presence

Detecting Medical Reviews

(FP). FF is the number of occurrences in the document. TF-IDF is given by TF-IDF = FF×Log(N/DF), where N indicates the number of documents, and DF is the number of documents that contain this feature. FP takes the value 0 or 1 on the basis of features absent or present in the document file.

11.3 DATA TRANSFORMATION

In this step, first the HTML tags are removed. Then the abbreviations are expanded using pattern recognition and regular expression techniques, and then non-alphabetic signs are removed. Then stop words are removed using a stop list constructed from various standard stop lists, with some changes being made related to the characteristics of the text.

For rejection, first we tag the negated word with the following words till the first punctuation mark occurrence. The use of this unigram is as a classifier. The results were compared before and after adding the classifier and no significant difference was found between them. The conclusion was found to be consistent with the findings. The reason for this is that it is hard to find a match between the tagged negation phrases among the whole set of documents. For this particular reason, after the negation, we reduced the tagged words to three and then to two because of the syntactic position. Due to this, more negation phrases could be added as the unigrams in the final set of reduced features. In addition, a reduction in redundancy stemming was performed on the documents. The number of features was reduced from 10,450 to 7,614 in Dat-1400, and in Dat-2000 it was reduced from 12,860 to 9,058 features.

After the above process, for the three different features weighting—FF, TF-IDF, and FP—three feature matrices were constructed for each of the datasets. To make clear, in the FF matrix, the $(i,j)^{th}$ entry is the FF weight of feature i in document j. Sets of experiments were carried out on the feature matrices of Dat-1400.

11.4 FILTERING

Filtering is being carried out using the univariate chi-squared method. For computing the dependency between the word and the category of text it is mentioned in, the chi-square method is used. Chi-square value is low if the word is frequent in many categories, while if the chi-square value is high, it means the word is frequent in less categories.

11.5 LEXICON-BASED METHOD OF SENTIMENT ANALYSIS: LEXICON CONSTRUCTION

According to Grabner *et al.* (2012), the entry in the lexicon is used to derive the overall sentiment value of the text, and in the lexicon-based sentiment analysis, quality is the main issue. The lexicon is generated on the basis of the vocabulary in the training set only. The token and part-of-speech (POS) tag is used for deriving entry in the lexicon. There is no additional source of data being used for the construction of the lexicon. This makes the lexicon highly domain specific, which is useful for sentiment analysis. The lexicon contains tokens, with each token assigned a sentiment value. Values greater than zero denote positivity, values less than zero denote negativity, and zero value denotes neutrality. For example, in the case of reviews about hotels, token "dirty" shows negativity while the token "beautiful" shows positivity. The assignment of sentiment values to the tokens in the lexicon is done prior to the classification. For the construction of the lexicon, neuro-linguistic programming (NLP)-based meta-data is used which selects the relevant subset of tokens. The processing steps include tokenizer and POS tagger. The tokenizer part identifies the relevant lexical units, while the POS part assigns the word category to each token. To make the size of the lexicon large, all the verbs and nouns are considered relevant. Table 11.1 shows the most frequent tokens extracted from the training set of hotel reviews.

For correct classification purposes, only those tokens that discriminate between different class labels are considered. Thus, for each class label, a separate lexicon is constructed which contains all the characteristic tokens of that class. The metric used is the relative frequency of the token with respect to POS tag and class label. If the relative frequency of the token for that class label is higher than the relative frequency of the same label for other class labels, only then is a token relevant for the class. An additional parameter α can be used to control the size of the lexicon. If $\alpha = 0$, it means that each token is assigned to that class labels in which it occurred at least once. If $\alpha = 1$, each token is assigned to that class label with highest relative frequency. For $\alpha > 1$, each token is assigned to the class label with a relative frequency which is α times higher than the relative frequency of the same token in all other classes. Thus for higher values of α, only those tokens will remain that occur only once among the all the classes. This method ensures that all the lexicons are disjoint sets of tokens with $\alpha > 1$.

The experiments conducted in further sections will be conducted on the lexicons generated with $\alpha = 2$. It ensures that all the tokens specific for a class minimum amount of overlapping tokens are included for a class. For a broad range of applications, there will be a distinction between only three class labels: positive, neutral, and negative. Since our reviews feature a 5-star rating, we will classify 1- and 2-star reviews as negative reviews, 3-star reviews as neutral reviews, and 4- and 5-star reviews as positive reviews, as shown in Figure 11.1. Since we are concerned with

TABLE 11.1
Extraction of Tokens

Token Words	Frequency	POS
Breakfast	354	Noun
Stay	426	Verb
Location	510	Noun
Staff	618	Noun
Room	1852	Noun
Hotel	1851	Noun

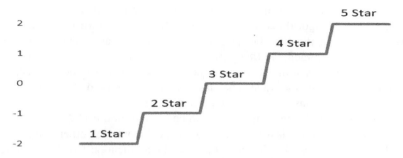

FIGURE 11.1 Sentiment analysis.

TABLE 11.2
Size of Lexicon

α	Positive	Negative
0	4327	5216
1	2809	3983
2	2384	3455
3	2240	3225
100	2058	2974

classes of positive and negative, this will require generating two dictionaries, as this will provide us with the best possible classification performance.

The addition of lexicons as neutral reduces other lexicons of terms that are referred to as "nearly positive or negative in sentiment," which leads to the reduction of the accuracy of output for this approach. Table 11.2 showcase the size of the lexicon in accordance with reduced labels.

A decrease in the number of class labels affects the size of the lexicon as the single training data sets are much larger, so we choose lexicons resulting in $\alpha = 2$ to further maintain comparability.

11.6 CLASSIFICATION

To calculate the sentiment score of any reviews, the lexicon journal is associated with other lexicons of similar orientation. All the entries in the distinct lexicons are given different values. It is assumed that any document which is without any given values identifies as neutral and thus the sentiment score given is 0. Hence no prior probabilities for reviews to belong to certain class are considered. Since we are using the 5-star rating system, the values associated are −2 given to every 1-star review, −1 to every 2-star review, 0 to every 3-star review, +1 to every 4-star review, and +2 to every 5-star review, as shown in Figure 11.1.

The classification fn. is computed on the basis of a sentiment analysis of the documents. During training, data has been treated and the lexicon generated using it.

To do sentiment analysis of the text, the sum of all the identified sentiments is calculated. Advanced algorithms are not necessary because here we are not accounting for negation and clarification. For performing sentiment classifying, the average of all the values of documents in that particular section label is used. This function is used to calculate the sentiment score of the text and then calculate the distance of that score from the class average. The class having minimum distance to that score of text is used as the class label for that particular text/review.

Sentiment values of three class labels can be seen in Figure 11.2.

Functions of different class labels are so different from each other that characteristics of the class are well reflected. The second classification which computes for three classes works the same way. The entries for the lexicon with a positive class label are assigned a value of +2, and entries for the lexicon with a negative class label are assigned a value of −2.

The higher distance between two different classes is because of the increased distance between their sentiment score in the dictionary, which has a large number of entries.

11.7 EVALUATION

Before computation of classification, the data set was divided into test and training sets containing 90% and 10% of the data respectively for each given class label. Then various different test sets containing multiple documents are evaluated separately. Evaluation metrics is used as recall and precision measures. *Precision* indicates the proportion of all reviews classified correctly to all the records classified, and *recall* is described as the portion of reviews selected correctly to sum of reviews selected. An F measure is also used, which combines both recall and precision into a single measure by calculating the harmonic mean. This F measure uses parameters to control the influence of precision and recall. As currently there is no particular application, the precision and recall have been set as equally important, and thus the weights given to both are 0.5. The performance is compared with baseline which is done by setting class labels randomly. Precision, recall, and F value based on 5 ratings and polarity are shown in Figure 11.3 and Figure 11.4.

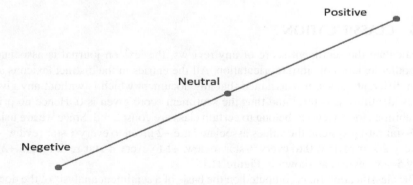

FIGURE 11.2 Sentiment values of three class labels.

Detecting Medical Reviews

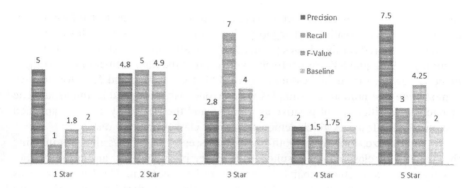

FIGURE 11.3 Higher distance between two different classes.

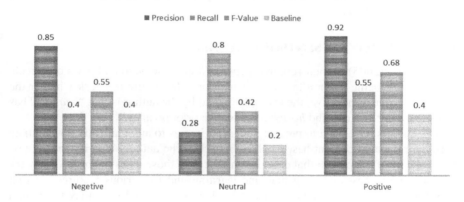

FIGURE 11.4 Precision, recall, and F-value of three different classes.

The precision, recall, and F-value for all classes is extensively higher than the benchmarked value. A more intensive examination of all these class qualities uncovers a few facts. The fact that the review for the 1-star and 5-star surveys is exceptionally low is because high assumption values are required for the report to be named effectively. The high score is accomplished either because of an enormous number of positive or negative conclusion-bearing tokens in the content. This could be because either the preparation set is excessively small or it is not specific enough to experience the audits that are not part of the preparation set. Then again, the accuracy value for 5-star and 1-star audits is high in light of the fact that not many surveys from different classes are scored that fall into this classification. The 4-star class is a negative exception on the grounds that 70% of the audits in this classification are inaccurately named 3-star surveys. Again, this demonstrates that not very many semantically pertinent tokens were discovered, which suggests that there a need to improve the nature of the dictionary. A conclusion score of 0 demonstrates that the audit is either nonpartisan or there are no semantic markers in the survey. This again proceeds to demonstrate the shortcomings of the main utilized vocabulary.

Next, the assessment was carried using the second technique for the grouping, where the three class names were utilized. For the word reference process, the surveys of the various classes were combined so the size of the test informational collection for both positive and negative class names comprises 40 audits each. The outcomes acquired for the classes positive and negative are almost comparable, with accuracy of 92% and 84% respectively, which shows the high caliber of the arrangement. Likewise, the review estimations of 40% for negative and 55% for positive demonstrate that numerous examples were erroneously delegated as impartial. The erroneous classification of positive as negative and the other way around happened just once in the test set, which demonstrates that classes are very distinct.

To summarize, looking at both displayed assessments shows that diminishing the quantity of class names resulted in a bigger preparing set, and a bigger dictionary would altogether expand the general order execution. The framework was likewise assessed with one vocabulary for every objective mark; however, the result was mediocre compared to the present arrangement utilizing two vocabularies for negative and positive.

11.8 A REVIEW OF SENTIMENT ANALYSIS

Wilson *et al.* (2005) in their research paper talk about lexicon and parts of speech, with lexicons being given a tag with their prior polar nature to check whether the word is positive or negative; the example quoted by the author is that "*beautiful* has a positive prior polarity and *horrid* has negative prior polarity."

The authors have tried to present new experiments to automatically differentiate between prior and content-based polarity checks. The authors identify the contextual polarity of the phrase that contains instances of those clues in corpus. They try various approaches, including machine learning and its various features. At first they classify the clues as neutral or polar. Then, in a second step, they try to find the polarity of those clues which were identified in step one, distinguishing them as positive, negative, or neutral.

They also developed a manual annotation scheme to determine the contextual polarity of all the subjective expressions. This will assign the polar value of the subjective equation as positive, negative, or both, or even neutral. Positive tags show positive emotions, evaluation, or stances. Negative tags show negative emotions, evaluation, or stances. A neutral tag is used for all the subjective expressions.

They conducted an agreement study and also applied a corpus model to check the subjectivity of words by dividing data into two sets, one for testing and one for training. Then the authors tried to find the prior polarity of subjective lexicon. They used more than 8,000 subjectivity clues. Basically, these are the words or phrases that can be used to express private states, and for this a single word clue is used.

They tried to compile and group the list of subjectivity clues on the basis of reliability as subjective clue, dividing them into two parts: strongly subjective and weakly subjective. They found that the earlier model was able to predict up to 82%, but the new model was able to predict up to 92.8%.

When they investigated the model further, they found that system was giving a label to each clue instance rather than identifying the system boundaries. Solving this problem can further improve the utility of the research. So, they tried to create a confusion matrix to check polarity on development set, and adopted a two-step approach to see the contextual sentiment and polarity disambiguation. So for the basic step, the focus was on whether hint instances are neutral or polar in context, and in the next step, they took all the clues instances marked as polar in the first step and focused on identifying their contextual polarity.

Finally, it is concluded that phrase-level sentiment analysis helps us to understand the basic type of statement, that is, whether it is neutral or polar; if it is polar then the authors tried to disambiguate the statement. So from this they were able to automatically identify the contextual polarity for a large data set of multiple expressions, and they were able to achieve better results as compared to the earlier model.

Singh *et al.* (2013) used different tools like machine learning and algorithm-based text classification (like the Naïve Bayes method, SVM, and kNN) along with a relevant features selection scheme. They also adopted an unsupervised semantic orientation scheme for extraction of relevant data, and used the Senti-Word-Net publicly available library which provides a positive, negative, or neutral score for words. This Senti-Word-Net approach helps us to get the sentimental score of every statement by looking up the terms from the text in the library.

The authors focus on "adjectives" and "adverbs" and also gave weights to different verbs, adverbs, and adjectives depending upon certain conditions; for example, an adverb should modify the sentimental score of the succeeding adjective or verb. To achieve greater accuracy, they took the modified weight (scaling factor) of the adverb score as 0.35.

We can see that if the adjective score is 0 then we ignore it; when the adjective is positive and the weights of adjective score is greater than 0, then we choose a minimum out of (1, score of the adjective added to the product of the scaling factor and the adverb score); if the adjective is negative, then we choose the maximum number out of (–1, score of adjective added to the product of scaling factor and score of adverb); and in the end we find the final sentiment value.

In the second step, when verbs are scored 0 then we ignore them; when an adjective is positive and a verb is greater than 0, then we choose minimum out of (1, score of the verb added to the product of the scaling factor and the score of the adverb); similarly, when the adjective is negative, then we choose maximum of (–1, score of the verb added to the product of the scaling factor and the score of adverb); also, if the verb is less than 0, then we choose the maximum out of (–1, score of the verb subtracted from the product of the scaling factor and the score of the adverb), and in the end we get the final Senti-Word-Net value.

This aspect-level analysis for sentiments involves various things like identifying which aspect is to be analyzed, then locating the opinionated content of that particular aspect in that review, and then finally determining the polarity of the view expressed about that opinion. The authors collected reviews from various sources and tried analyzing them, and from this model they were able to generate accuracy to great extent.

They also tested their model on Alchemy API, an IBM Watson software, to check polarity, and it was found that the model accuracy was even better than Alchemy API; the authors' model is able to give better results, as in data set of 760 positive reviews, it was able to calculate up to 688, while Alchemy API was only able to calculate 634 positive reviews.

From this research paper, it can be concluded that firstly they explored the use of adverbs and verbs combined with adverbs and adjectives for document-level sentimental classification. Secondly, the research was based on the features of the heuristics scheme for a basic level of polarity classification of words. And thirdly, it resulted in congruent results between aspect-level and document-level sentiment classification.

Kasper and Vela (2011) have developed their own process of text classification and sentimental analysis for hotels called BESAHOT. This service provides a helping hand to hotel managers, collecting user reviews from various different websites, analyzing them and classifying the textual form of review, and presenting the output in a concise manner.

Their model presents a system that automatically monitors user comments on the Web from various sites and provides classified summaries of positive and negative features of a hotel. After this comes content extraction and duplicate checking; in the extraction step, it collects data from all the websites listed and removes opinions given by the same person from different sources. Then comes the internal process of BESAHOT, which performs various steps pf sentiment analysis, finds a joint polarity, and gives a final rating to a hotel. From this we are able to get the joined results of many people at the same time; normally, some hotels can be rated good on one website and bad on another website, so this helps us to arrive at a common answer from combined opinions.

When we investigate this paper further, we find that this model was able to generate good results as the values in the final results were able to predict up to 90% (approx.) which shows the model is working excellently, as the authors have also tried to code these words by doing stemming and getting words to their root words. The authors also say that they used already existing polarity checking tools. Here we can conclude that this is one way of performing opinion mining and sentimental analysis by considering data from different websites available on the web, but the authors think that their model can be made even more useful by adding multi-topic segmentation to give a better interactive user interface.

Leung *et al.* (2008) conveyed that there are various levels of doing sentiment analysis, like word level analysis, sentence analysis, and even document-level analysis. According to them, most existing sentiment-analysis algorithms were designed for classifying in only binary terms, which means that they will assign opinions or reviews to a bi-polar class of being either positive or negative. Some researchers have taken this study to the next level by grading this sentiment analysis on a multipoint rating scale, for example, on a scale of 1 to 5. They also tried to link sentiment analysis to sentiment summarization level, but according to the authors, subjectivity analysis was closely related to sentiment analysis because that determines the nature of text, and for this analysis they classified it in two different ways: a subjective and objective class, and a parts-of-speech class. This was the basic layout identified by the authors for sentiment analysis; from this it can be inferred that firstly they collected the reviews and tried to present the data, and then did the basic analysis for reviews and tried to find out whether the review was positive or negative, and prepared the sentiment analysis test to check the polarity of the review. For this they tried to use a Sentimental Orientation (SO) approach and a machine learning approach (Figure 11.5).

Detecting Medical Reviews

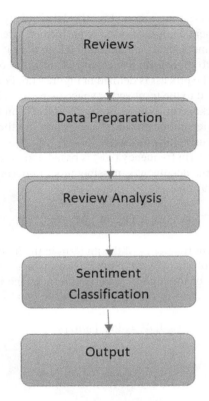

FIGURE: 11.5 A typical sentiment analysis.

From this study, we can conclude that sentimental analysis basically deals in classifying textual data based on the sentiments that it contains. Sentimental analysis has emerged as a trend for research in the areas of text mining and computation of linguistics. Future research shall explore more sophisticated methods of opinion mining and product feature extraction, as well as newly classified models that can be addressed by properly ordered labels in the rating of inferences.

11.9 PERFORMING SENTIMENT ANALYSIS ON MEDICAL REVIEWS

Sentiment analysis was carried out on doctor reviews found online on various websites like mouthshut.com. The sentiment analysis was carried out using the Python programming language's Sklearn package. The first process involved importing the data so that it could be used for the analysis purpose. After the data is imported, the first step is to clean the data of unwanted text, so that text which add no sentiment value to the data is removed. The cleaning involves the removal of special characters and stop words, bringing the words back to their root form.

After the process of cleaning is complete, the data is ready to be used for identifying sentiments. Sentiment analysis cannot be performed directly on the textual data as computers cannot understand human language, and so it becomes necessary to convert textual data into some form that can be used by the computer to compute the sentiment of the text. The process is defined in Figure 11.6. For the same purpose, the textual data is converted to the term vector space using the TF-IDF vectorizer function of the Sklearn package of Python. In the TF-IDF (Term Frequency and Inverse Document Frequency) technique, the term vector space for the reviews is made using the weights calculated using this method.

The concept behind term frequency (TF) is based upon the fact that a term which occurs many times in the document is possibly more important than the term which occurs lesser number of times.

$$TF_{ij} = F_{ij}/M_i,$$

where TF_{ij} is the term frequency of the jth term in the ith document.
F_{ij} is the frequency of the jth term in the ith document.
M_i is the frequency of the highest occurring term in the ith document, i.e. $M_i = \max(F_{ij})$

$$IDF_j = \log_2(n/n_j) + 1,$$

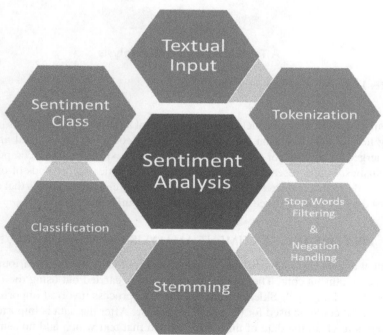

FIGURE 11.6 Performing sentiment analysis on medical reviews.

Detecting Medical Reviews

where n is the total number of documents and n_j signifies the number of documents in which the jth term occurs.

The concept behind inverse document frequency (IDF) is that a term that occurs in fewer documents is likely to be a better discriminator than a term which occurs in most or all documents.

Using the TF-IDF method, the weights are calculated for the term vector space of reviews. The TF-IDF is calculated for each term, and this value of TF-IDF is used as the weight in the term vector space. Every review is converted to the term vector space, which is then further used for the purpose of sentiment analysis. This term vector space model is used for the purpose of machine learning as it makes it easier for the machine to identify the sentiment based on the weights of certain terms.

After the text has been converted to the term vector space using the TF-IDF method, the dataset of reviews is divided into a training set and a testing set. The machine learning tool which we are using here is the Random Forest classifier. We have used the Sklearn ensemble library for importing the Random Forest classifier, which is a machine learning algorithm.

This machine learning model is trained using the training dataset of the term vector form of the reviews. The training dataset contains the reviews in the form of term vector space and their corresponding sentiment value. The machine learning model uses this training data and its sentiment values to construct the model. The model created is then tested on the testing data to check its accuracy. During testing, the model is provided with unseen data, and the model predicts the sentiment of the reviews. The predicted value of the sentiment is then matched with the actual values of the sentiment of the reviews. A higher accuracy means that the model created is good.

The whole dataset is split into the training and testing dataset in a ratio of 70% and 30% ratio respectively. This split is carried out using the train_test_split function of the Sklearn library. It randomly picks the reviews from the dataset and then arranges them into training and testing sets.

The accuracy of any model depends on the dataset which is being used for training. The higher the number of data used to train the model, the higher the accuracy of the model created. The higher the precision value of the model, the better it will predict the sentiments of unseen datasets.

So, from this we can see that the output is significant (Table 11.3) as it is significant enough to predict the right number of cases.

TABLE 11.3
Output of Medical Reviews

			Predicted		
			Positive	Negative	Neutral
			871	377	252
Actual	Positive	780	614	79	87
	Negative	492	157	225	110
	Neutral	228	100	73	55

11.10 SENTIMENT ANALYSIS WORKING PROGRAM

Training data is used to help the computer learn how to deal with information; by using algorithms, the computer is trained to mimic the human brain so as to receive the input, analyze the input, and then give the desired result. This is usually done manually and it follows the same rules and instructions given to it. To train the model, the following are needed:

- Input training dataset
- Name of the data attribute that contains the target to be predicted
- Instructions for required data transformation
- Training parameters required, if any, so as to control the learning algorithm

Out of 8,000 tweets captured, a data set of 800 was shortlisted, and from that, 720 tweets were used for training and 80 tweets were used for testing.

For this data, an excel file was created which contains two columns: Tweet and Sentiment. The tweet column contains the actual Tweets which were there in the excel file which was exported from the NVivo application, and the sentiment column contains the sentiment value written in the form of classes, that is, positive, negative, and neutral.

The classes are written in the form of numeric representation where:

0 = positive sentiment
1 = negative sentiment
2 = neutral sentiment

The excel file was created for the purpose of training our sentiment analysis model. The column highlighted in blue contains the actual tweets which were extracted from Twitter.com using the NCapture extension. The column highlighted in yellow contains the corresponding sentiment value.

Now this excel file will be used for training the sentiment analysis model and also for testing it.

The model for the sentiment analysis has been made using the Random Forest classifier method of the Sklearn library of the Python.

The training data which have been created above will be used to train the Random Forest classifier model.

In the code shown in Figure 11.7, the underlined code at the top shows the method of importing the Random Forest classifier method of the Sklearn library of the Python programming language. The underlined code at the bottom shows use of this Random Forest classifier model. The "n_estimators = 1000" parameter of the function tells the computer to make 1,000 decision trees in the Random Forest model.

Since computers do not understand data in text format, it is necessary to convert all the Tweets in the training data file into a format which will be understood by the computer. For the purpose of converting the Tweets into the format which will be easily interpreted by the machine the TF-IDF method has been used. To use this

Detecting Medical Reviews 213

```
15 from sklearn.ensemble import RandomForestClassifier
16 from sklearn.metrics import accuracy_score
17 import nltk
18 nltk.download("stopwords")
19 from nltk.corpus import stopwords
20
21 stemmer = WordNetLemmatizer()
22
23 tweet_file = pandas.read_excel("./datasets/data101.xlsx")
24
25 tweet_data = tweet_file["tweet"].values
26
27 y = tweet_file["sentiment"].values
28 tweets = []
29 for z in tweet_data:
30     t = ' '.join(re.sub("(@[A-Za-z0-9]+)|([^0-9A-Za-z \t])|(\w+:\/\/\S+)", " ", z).split())
31     t = re.sub("\s+[a-zA-Z]\s+"," ", t)
32     t = t.lower()
33     t = t.split()
34     t = [stemmer.lemmatize(word) for word in t]
35     t = " ".join(t)
36     tweets.append(t)
37
38 #print(tweets)
39 #print(sentiment)
40
41 tfidfvectorizer = TfidfVectorizer(max_features = 3000, min_df = 2, max_df = 0.9, stop_words =
42 x = tfidfvectorizer.fit_transform(tweets).toarray()
43 print(len(x[0]))
44 x_train, x_test, y_train, y_test = train_test_split(x, y, test_size = 0.1, random_state = 28)
45
46 classifier = RandomForestClassifier(n_estimators = 1000, random_state = 0)
47 classifier.fit(x_train, y_train)
```

FIGURE 11.7 Random Forest classifier.

method in the project, the TF-IDF Vectorizer method of the Sklearn library has been used. This method converts all the tweets present in the training excel file into a term vector space which will be easier for the machine to use as machines cannot understand human languages.

In the code Figure 11.8, the underlined code at the top shows how the TF-IDF vectorizer method is imported in our code. The underlined code near the bottom shows the usage of the TF-IDF Vectorizer method, the parameter "max_features = 3000" tells the machine to make at most 3,000 terms for the term vector space, which means that the term vector space will be made up of no more than 3,000 terms.

The parameter "min_df = 2" means that to use terms which are present in at least two Tweets and the parameter "max_df = 0.9" means to use the terms that are present in at most 90% of all documents.

The parameter "stop_words = stopwords.words('english')" tells the machine to neglect the stop words for making the term vector space. The list of stop words are downloaded using the NLTK (natural language tool kit) library of the python. The NLTK library contains the stop words for all languages. In this case, the stop words list for the English language was required.

The underlined code at the bottom converts the term vector space into a Python array. The array contains the term vector space in the form of Python arrays.

For training the model, the training data is needed. In the above code, the training data has been converted into the term vector space. Now the Random Forest classifier model will be trained using this data. To train and test the model, the training data is split into the training set and test set.

```
13 from sklearn.feature_extraction.text import TfidfVectorizer
14 from sklearn.model_selection import train_test_split
15 from sklearn.ensemble import RandomForestClassifier
16 from sklearn.metrics import accuracy_score
17 import nltk
18 nltk.download("stopwords")
19 from nltk.corpus import stopwords
20
21 stemmer = WordNetLemmatizer()
22
23 tweet_file = pandas.read_excel("./datasets/data101.xlsx")
24
25 tweet_data = tweet_file["tweet"].values
26
27 y = tweet_file["sentiment"].values
28 tweets = []
29 for z in tweet_data:
30     t = ' '.join(re.sub("(@[A-Za-z0-9]+)|([^0-9A-Za-z \t])|(\w+:\/\/\S+)", " ", z).split())
31     t = re.sub("\s+[a-zA-Z]\s+"," ", t)
32     t = t.lower()
33     t = t.split()
34     t = [stemmer.lemmatize(word) for word in t]
35     t = " ".join(t)
36     tweets.append(t)
37
38 #print(tweets)
39 #print(sentiment)
40
41 tfidfvectorizer = TfidfVectorizer(max_features = 3000, min_df = 2, max_df = 0.9, stop_words = stopwords.words("english"))
42 x = tfidfvectorizer.fit_transform(tweets).toarray()
```

FIGURE 11.8 TF-IDF Vectorizer.

To do this, the train_test_split method of the Sklearn library is used. In the code Figure 11.9, the part underlined at the top shows how the train_test_split method of the Sklearn library is imported.

The part underlined at the bottom shows training data being split into 90% – 10% parts. 90% of the data is used for training the Random Forest classifier model and 10% of the data is used for testing the trained model. The parameter "test_size = 0.1" tells the machine to use the 90% of data as the training set and 10% of the data as the test set. After training is done, the model created using the training dataset is tested on the test dataset, which is 10% of the data.

After the trained model is run on 10% of the test data, the accuracy achieved is 75% (Figure 11.10), and two excel files are saved in which one file contains positive tweets and one file contains negative tweets.

11.11 CONCLUSION

From all the research papers discussed, it can be concluded that there are various ways to determine sentiments in data, whether on the basis of POS, word level analysis of sentences, or document level analysis of sentiments, and various work is ongoing to improve model accuracy. The results of one of the papers shows that when we do pre-processing we are able to generate better results as compared to without a pre-processing model. We can also create different classes for the data to have a better understanding of the model and also improve the accuracy of the results.

Detecting Medical Reviews 215

```
14 from sklearn.model_selection import train_test_split
15 from sklearn.ensemble import RandomForestClassifier
16 from sklearn.metrics import accuracy_score
17 import nltk
18 nltk.download("stopwords")
19 from nltk.corpus import stopwords
20
21 stemmer = WordNetLemmatizer()
22
23 tweet_file = pandas.read_excel("./datasets/data101.xlsx")
24
25 tweet_data = tweet_file["tweet"].values
26
27 y = tweet_file["sentiment"].values
28 tweets = []
29 for z in tweet_data:
30     t = ' '.join(re.sub("(@[A-Za-z0-9]+)|([^0-9A-Za-z \t])|(\w+:\/\/\S+)", " ", z).split())
31     t = re.sub("\s+[a-zA-Z]\s+", " ", t)
32     t = t.lower()
33     t = t.split()
34     t = [stemmer.lemmatize(word) for word in t]
35     t = " ".join(t)
36     tweets.append(t)
37
38 #print(tweets)
39 #print(sentiment)
40
41 tfidfvectorizer = TfidfVectorizer(max_features = 3000, min_df = 2, max_df = 0.9, stop_words = stopwords.words("english"))
42 x = tfidfvectorizer.fit_transform(tweets).toarray()
43 print(len(x[0]))
44 x_train, x_test, y_train, y_test = train_test_split(x, y, test_size = 0.1, random_state = 28)
45
```

FIGURE 11.9 Train_test_split method of Sklearn library.

```
[2 2 1 1 2 1 0 0 0 2 1 1 0 1 0 2 1 1 0 2 1 1 2 0 1 0 2 2 1 2 1 1 1 0 0 0
 0 2 2]
[2 2 0 2 2 1 2 0 0 0 1 1 1 0 1 0 2 2 1 2 2 1 1 2 0 1 0 1 2 1 2 1 1 1 1 2 2
 0 2 2]
Accuracy:
75.0
```

FIGURE 11.10 Achievement of accuracy.

    The sentiment analysis can be used by industry to know the perception of their products and services in the market. It can be used by companies to gather knowledge of how well their products and services are performing in the market and how the customers view them. With sentiment analysis, hospitals can know how the services provided by them are perceived by the customers. It can be used to determine the positive and negative points of their services, which can be used to improve their services. Sentiment analysis can also be used to find negative reviews which will help in pointing out the negative points about the services in hospital which will in turn help in improving the services by dealing with those negative points. Similarly this process can be used for medicines, doctors, and health equipment.

BIBLIOGRAPHY

Haddi, Emma, Liu, Xiaohui, and Shi, Yong. 2013. The role of text pre-processing in sentiment analysis, Beijing. *Procedia Computer Science*, 17, 26–32.

Grabner, Dietmer, Zanker, Markus, Fliedl, Gunther, and Fuchs, Matthias. 2012. Classification of customer review based on sentiment analysis, Sweden. *19th Conference on Information and Communication Technology in Tourism*, Nicosia, Cyprus.

Wilson, Theresa, Wiebe, Janyce, and Paul, Hoffmann. 2005, October. Recognizing contextual polarity in phrase-level sentiment analysis. *Proceedings of the Human Language Technology Conference and Conference on Empirical Methods in NLP (HLT/EMNLP)*, Vancouver.

Singh, V. K., Piryani, R., Uddin, A., and Walia, P. 2013, March. Sentiment analysis of movie reviews, India. *Conference Paper*, Delhi and Varanasi, India. doi:10.1109/imac4s.2013.6526500.

Aue, A., and Gamon, M. 2005. Customizing sentiment classifiers to new domains: A case study. *Proceedings of the Recent Advances in Natural Language Processing*, Bulgaria.

Blitzer, J., Dredze, M., and Pereira, F. Biographies, Bollywood, boom-boxes and blenders: Domain adaptation for sentiment classification. *Proceedings of the 45th Annual Meeting of the ACL*. Retrieved June 23, 2007 from http://acl.ldc.upenn.edu/P/P07/P07-1056.pdf.

Bruce, R., and Wiebe, J. 1999. Recognizing subjectivity: A case study of manual tagging. *Natural Language Engineering*, 5(2), 187–205.

Dave, K., Lawrence, S., and Pennock, D. M. 2003. Mining the peanut gallery: Opinion extraction and semantic classification of product reviews. *Proceedings of the 12th International World Wide Web Conference*, Budapest, Hungary.

Esuli, A., and Sebastiani, F. 2005. Determining the semantic orientation of terms through gloss classification. *Proceedings of the ACM Conference on Information and Knowledge Management*.

Gamon, M., Aue, A., Corston-Oliver, S., and Ringger, E. 2005. Pulse: Mining customer opinions from free text. *Lecture Notes in Computer Science*, 3646, 121–132.

Goldberg, A. B., and Zhu, X. 2006. Seeing stars when there aren't many stars: Graph-based semi-supervised learning for sentiment categorization. *Proceedings of the TextGraphs Workshop*, Vancouver, Canada.

Hu, M., and Liu, B. 2004. Mining and summarizing customer reviews. *Proceedings of the 10th ACM SIGKDD International Conference on Knowledge Discovery and Data Mining*, California, USA.

Hu, M., and Liu, B. 2004. Mining opinion features in customer reviews. *Proceedings of the 19th National Conference on Artificial Intelligence*, pp. 755–760, California, USA.

Kasper, Walter, and Mihaela, Vela. 2011. Sentiment analysis for hotel reviews. *Proceedings of the Computational Linguistics-Applications Conference*, pp. 45–52.

Leung, W. K. Cane, and Chan, C. F. Stephen. 2009. *Sentiment Analysis of Product Reviews*, Department of Computing, The Hong Kong Polytechnic University Hung Hom, Kowloon Hong Kong SAR.

Index

A

Accelerometer, 66
Algorithm, 185
Analytics, 184
Antagonism, 64
Application programming interface (API), 14, 124
Attribute-based access control (ABAC), 126
Arbitrarily, 69
Artificial intelligence (AI), 32
Assistive, 186
Authentication, 144-146, 148, 152

B

Behavioral intention (BI), 69
Bharat Interface for Money(BHIM), 62
Big data, 40, 135–146, 152
Big—Data, Analytics, and Decisions (B-DAD), 136
Bluetooth, 118
    controller, 67
Brute force attacks, 106
Business-to-business (B2B), 58
Business-to-consumer (B2C), 58

C

Certificate authority (CA), 148
Chi-Square, 51–56, 200–201
Chronic, 83
Cloud, 3, 4, 5, 13
    computing, 39, 85, 86
Classification, 141, 200, 202–208
Clustering, 141–143
Compound annual growth rate (CAGR), 6
Computer-aided detection (CADe), 82
Computerized tomography (CT), 79, 81
Community of interest (COI), 150
Computer vision, 184
Consumer privacy, 126
Convolution neural network (CNN), 80
Constrained application protocol (CoAP), 147
Critical success factor (CSF), 164
Cryptography, 127
Cronbach's alpha, 167, 175
Cyber espionage, 106

D

Data investigation, 128
Data processing, 136, 138, 139, 143
Data protection, 126
Data transformation, 201, 212
Decision support systems (DSS), 80, 91
Deep learning, 78–81, 82, 85
Denial of service (DoS), 105–107, 112
Deoxyribonucleic acid (DNA), 59
Desktop vocational assistant robot (DeVAR), 186
Dictionary attack, 107
Digital drug, 59
Domain name system (DNS), 148
Domain Name Service Security Extension (DNSSEC), 148
Drive and Share (DaS), 152–153
Drones, 190

E

Eavesdropper, 104
Electrocardiogram (ECG), 35, 37, 38, 80, 86
Eigenvalue, 178
Emerging, 189
Encryption, 104–105, 108, 146–147, 149
Electronic health records, 73
Essential administrations, 126
Exploratory factor analysis, 167
Equilibrium, 187

F

Factor analysis, 175, 176, 178
Feature frequency, 201
Feature presence, 200–201
Filtering, 200–201
Food and Drug Administration (FDA), 50
Frequency, 200, 202, 210, 211
F-value, 204–205
Fusion, 188

G

5G, 6, 7
Gadgets, 125–127
General Data Protection Regulation (GDPR), 122
Global navigation satellite systems (GNSS), 7

H

Harmonization, 127
Hadoop Distributed File System, 139–140
Heating, ventilation, and air-conditioning (HVAC), 18

Index

Healthcare, 34, 35, 38, 39, 40, 41
   device, 98
Hewlett-Packard (HP), 28
Human-to-human interaction (HHI), 23
Human-to-computer interaction (HCI), 23
Hyperthyroidism, 67
Hypothesis testing, 169
HTML, 200–201

I

Imitation, 185
Impassable, 68
Industrial Internet of Things (IIoT), 30
Ingestible sensor, 58
InstaMed, 68
Interoperability, 57
Information technology (IT), 165
Information and communication technology (ICT), 32
Intel Health Application Platform (IHAP), 37
Intensive care unit (ICU), 36
Internet of Medical Things (IoMT), 85, 86
Internet of Things (IoT), 1, 2, 3, 4, 18, 23–32, 41, 42, 45–60, 85, 89, 94–103, 115–117, 136–139
   security, 120
Intrusions, 111
Inverse document frequency (IDF), 200–201, 210–212

K

Knowledge-based Augmented Reality for Maintenance Assistance (KARMA), 4

L

Lesions, 82
Lexicon, 200–206
Level of notoriety, 127
Likelihood ratio, 51–56

M

Machine learning (ML), 184, 199, 200, 206, 207, 208, 211
Machine-to-machine (M2M), 2, 24, 25, 26
Magnetic resonance imaging (MRI), 79–84, 91
Malicious attack, 94
Mammography, 82
Man-in-the-middle attacks (MITM), 100
Matrix, 201, 206
Microelectromechanical systems (MEMS), 24
Multiple regression, 167, 170
Multi-agent, 186, 194
Multi-dimensional, 187

N

Naïve Bayes, 200, 207
Natural language processing (NLP), 89, 202
Natural language tool kit (NLTK), 213
Near field communication (NFC), 47, 67
Next release problem (NRP), 164–165, 167
Nodules, 82
Non-repudiation, 107, 110

O

Onboard unit (OBU), 5
Oncology, 82

P

Parsimonious, 63
Part-of-speech (POS), 201–203, 215
Patient position device, 66
Polarity, 199–200, 204, 206–208
Proposed architecture, 66
Perceived convenience, 65
Personal computer (PC), 127
Personal digital assistant (PDA), 118
Personal emergency response system (PERS), 49
Positron emission tomography (PET), 81
Positron emission tomography-computerized tomography (PET-CT), 84
Precision, 204, 205, 211
Probability level, 173
Programmable logic controller (PLC), 121

Q

QR codes, 23
Quality of experience (QoE), 150
Quell, 37

R

Radio frequency identification (RFID), 2, 3, 4, 23, 24, 26, 31, 32, 46–47, 96, 105, 144–146, 154
Random Forest, 211–214
Raspberry Pi, 119
Recall, 204, 205
Recognition, 188
Reinforcement, 191
Reliability analysis, 164, 167
Research trends, 187
Robotics, 184

S

Scalability, 57
Science, technology, engineering, and mathematics (STEM), 184

Index

Security, 144–149, 154, 156
Self-supervised, 185
Semantic-directed vision, 47
Sentiment analysis, 199–212, 215
Sentimental Orientation (SO), 208
Sending layer, 48
Sklearn, 209–214
Smart cities, 153–155
Smart Tissue Autonomous Robot (STAR), 186
Sniffer, 106, 111–112
Social Devices Platform (SDP), 153–154
Social Internet of Things, 149–153
Social networks, 140, 149–153
Software development lifecycle (SDLC), 165
Subscriber identity module (SIM), 13
Supine, 66
Supervised, 190
Supply chain, 31, 37
Support Vector Machine (SVM), 200, 207
Supervisory control and data acquisition (SCADA), 24, 107
Standards of administration, 126
Statistical Package for the Social Sciences (SPSS), 51, 53, 54
Stratum, 69

T

Tachycardia, 67
Technology Acceptance Model (TAM), 63
Telematics, 5, 6, 14
Telemedicine, 39, 79, 85, 86
Term frequency (TF), 200–201, 210–212
Third-party administrators (TPA), 62
Transformation, 200
Transplantation, 59
Trojans, 107, 111
Trust framework, 128

U

Unconventionally, 64
Unmanned aerial vehicles (UAVs), 189
Unsupervised, 191
Unified Payment Interface (UPI), 62
Unique identifiers (UIDs), 23
Universal serial bus (USB), 26
Usage-based insurance (UBI), 14

V

Varimax rotation, 169
VSAT, 6
Vehicle-to-vehicle (V2V), 7

W

Warfarin, 63
Web of Data (WoD), 137–138
Wireless, 2, 5, 7, 9, 12
    patient observing, 49
Wireless sensor networks (WSN), 154
Wearable, 50, 85, 86
    sensors, 30, 33
World Health Organization (WHO), 9

Y

YuGo, 37

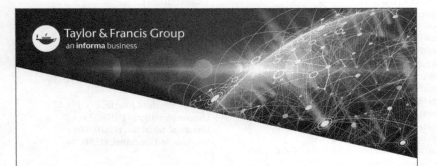